Fundamentals of
radiation chemistry

Fundamentals of radiation chemistry

A. R. Denaro, M.Sc., Ph.D., F.R.I.C.
G. G. Jayson, B.Sc., Ph.D., F.R.I.C.

LONDON   BUTTERWORTHS

CHEMISTRY

THE BUTTERWORTH GROUP

ENGLAND
Butterworth & Co (Publishers) Ltd
London: 88 Kingsway, WC2B 6AB

AUSTRALIA
Butterworth & Co (Australia) Ltd
Sydney: 586 Pacific Highway Chatswood, NSW 2067
Melbourne: 343 Little Collins Street, 3000
Brisbane: 240 Queen Street, 4000

CANADA
Butterworth & Co (Canada) Ltd
Toronto: 14 Curity Avenue, 374

NEW ZEALAND
Butterworth & Co (New Zealand) Ltd
Wellington: 26–28 Waring Taylor Street, 1

SOUTH AFRICA
Butterworth & Co (South Africa) (Pty) Ltd
Durban: 152–154 Gale Street

First published 1972

ISBN 0 408 70312 1 standard
     0 408 70313 X limp

Printed in England by Cox & Wyman Ltd,
London, Fakenham and Reading

Preface

Radiation chemistry is becoming more and more an integral part
of undergraduate courses, and this book has been written to provide
a first introduction to the subject. It is intended primarily for under-
graduates in the second year of their degree courses, when they have
already obtained some understanding of atomic and molecular
structure, together with the basic ideas of chemical kinetics. In view
of this, conventional kinetic techniques have not been discussed,
except where they have some specific application to radiation
chemistry.

The object of the book is to establish the basic principles under-
lying the behaviour of chemical systems exposed to ionising radiations. To give a reasonable understanding of the concepts which are
of fundamental importance to radiation chemistry, some space has
been devoted to the behaviour of ions, excited states and free radicals.
The chapter on water and aqueous solutions has also been used to
develop many of the general principles of radiation chemistry. At the
same time, the length of the book must be restricted to some extent
and it does not, therefore, aim to be a comprehensive survey of the
radiation chemistry of a large number of chemical systems. The
detailed radiation chemistry of many systems is still undergoing
development; for this reason only a few systems have been selected
to illustrate the general principles both in the field of aqueous solutions and in other areas. The reader is referred to the books listed in
the bibliography for more detailed accounts of a wider selection of
specific systems. In this way it is hoped that the student will gain an
insight into radiation chemistry without being overloaded by a vast
catalogue of facts at his first reading of the topic.

A final chapter has been added on some aspects of the radiation chemistry of biological systems. This is one of the main areas in which radiation chemistry finds application. It is desirable that radiation chemists should have some acquaintance with this field, both for the above reason and also to gain some understanding of the health hazards associated with ionising radiations.

A.R.D.

G.G.J.

Contents

1 Introduction   1
    Initiation of chemical reactions. Systems of units. The origin of radiation chemistry.

2 Sources of ionising radiations   13
    External sources. Internal sources. Radiation vessels. Glow discharge electrolysis.

3 The behaviour of ionising radiations   31
    Charged particles. Neutrons. Electromagnetic radiation. Effects of ionising radiations upon matter.

4 Behaviour of ions, excited states and free radicals   56
    Detection and study of ions. Reactions of ions. Reactions of electrons. Detection and study of excited states. Reactions of excited states. Detection and study of free radicals. Reactions of free radicals. Radiation chemical yields.

5 Dosimetry   82
    Ionisation dosimetry. Chemical dosimetry. Measurement of low doses. Dosimetry of internal sources. Neutron dosimetry.

6 Water and aqueous solutions   98
    The products of water radiolysis. The ionisation process. The excitation process. The mechanism of water radiolysis. Radical and molecular yields. Radiolysis of pure water. Limitations of the concepts of radical

and molecular yields. Radiolysis of aqueous solutions.
Determination of radical and molecular yields. Results
of radical and molecular yield determinations. Radiolysis of aqueous solutions at low concentrations. Determination of radical reaction rate constants. Reactions of
hydrated electrons. Reactions of hydrogen atoms.
Reactions of hydroxyl radicals. Defects of the diffusion
model of water radiolysis. Modifications to the diffusion
model.

7 Gases and solids   147
    Gases. Solids.

8 Organic systems   169
    Hydrocarbons. Polymerisation. Irradiation of solid
    polymers. Graft polymerisation. Halogen compounds.
    Alcohols. Acids, ethers, ketones and esters.

9 Aspects of biological systems   187

Bibliography   196

Index   198

1
Introduction

INITIATION OF CHEMICAL REACTIONS

Chemistry is concerned largely with the reactions occurring between chemical substances. These reactions can be considered as the breaking and making of chemical bonds with a consequent rearrangement of the atoms of the materials to form different molecules. A chemical reaction is thus initiated by the rupture of a chemical bond, and this in turn requires the input of energy into the system.

The ultimate feasibility of a spontaneous chemical reaction can be judged from thermodynamic criteria. If the chemical potential of the products is less than that of the reactants, the reaction is feasible. Whether the reaction proceeds spontaneously at a significant rate, however, is dictated by kinetic considerations. Only those molecules with an energy sufficient to overcome the activation energy barrier are capable of forming products. To provide such molecules, or to increase their number to the extent where the reaction proceeds at an appreciable rate, energy must be transferred to the system. Energy can be transferred to chemical systems in a variety of ways, and one of the commonest methods is simply to heat the system. Many reactions are thus accelerated or initiated by *thermal methods*. Conventional thermal methods usually achieve temperatures up to a few hundreds of degrees Celsius, but the more sophisticated technique of subjecting substances to *shock waves* can achieve temperatures of thousands of degrees at the front of the shock wave and high pressures.

A more selective energy input can be attained by *photochemical*

1

methods. Energy is deposited in the system by irradiating it with light. With wavelengths in the ultraviolet region, molecules can be excited to a higher electronic energy level and dissociated to form radicals. The radiation is absorbed selectively by those molecules which have excitation energies corresponding to the quantum energy of the wavelength of light used. Although this method is selective, the energy is usually deposited homogeneously throughout the system and the resultant chemical effects are due to the reactions of the energy-rich excited molecules and radicals.

Many reactions can be initiated in *electrical discharges*, but in these cases the situation is very complex and the products depend very much upon the nature of the discharge. In an electric discharge, electrons are accelerated by the electric field and impact with molecules to form excited states, ions and free radicals. As the electric field in a discharge is usually not uniform, the energies transferred to the system in various parts of the discharge differ, and the efficiency of a reaction can vary according to its location in the discharge.

It will be appreciated that photochemistry is selective, because it supplies just the right amount of energy needed to produce a required excitation process. In an electric discharge, however, ions are necessary to carry the current and enough energy must be imparted to a molecule by electron impact to ionise it. Moreover, the electrons in the discharge have a range of energies, so that the electric discharge technique will be non-selective in the transfer of energy which will be deposited inhomogeneously throughout the discharge.

Another method of initiating reactions is by *ultrasonics*. By subjecting a chemical system to very-high-frequency sound waves, small regions of high temperatures and pressures can be induced in the system. It is also believed that some ionisation is produced. Once again, excited states, free radicals and ions are generated, and the chemical effects are due to the reactions of these species.

Energy can also be transferred to chemical systems by *ionising radiations*. As the name implies, these are radiations which are capable of producing ionisation in any medium through which they pass. Such radiations are α-rays, β-rays and γ-rays arising from radioactive nuclei, and X-rays, which are electromagnetic radiation but of much shorter wavelength and, hence, greater energy than the wavelengths used in photochemistry. Electrons, protons, deuterons and other heavy positive ions, when accelerated, can also cause ionisation, and come into the category of ionising radiations. All the examples mentioned above can cause ionisation directly. Also included in ionising radiations are neutrons, which do not give rise directly to ionisation but usually produce fast-moving protons which subsequently cause ionisation. Slow neutrons may also produce radioactive material when they interact with matter.

It has already been pointed out that in photochemistry the choice of wavelength determines the quantum energy of the photons. For a large number of molecules the excitation energy corresponds to a wavelength in the ultraviolet region, and by choice of a particular wavelength the energy of the radiation will be specifically absorbed by a given molecule. Usually the whole energy of the photon is absorbed by one molecule, so that one photon can cause one excitation event. Thus the choice of wavelength will produce photons of just the right energy to cause a specific excitation in a particular type of molecule. If there are other types of molecule in the system, they may well be unaffected by the ultraviolet radiation. In contrast, the energy of ionising radiation is usually hundreds or thousands of times greater than the energies encountered in photochemistry. Ionising radiations thus have many times more energy than is required for the excitation or even the ionisation of one molecule. As a result, ionising radiations are not selective and may ionise or excite any molecule lying in their path. Moreover, as they start off with such high energies, they will interact with many molecules and will thus leave behind them a trail of ionised and excited molecules. The energy of ionising radiations is not absorbed uniformly by a medium, nor is it absorbed specifically by only certain molecules in the medium.

The contrast between the modes of energy deposition in photochemical methods and with ionising radiations is probably best appreciated from a consideration of solutions. The photochemistry of a substance may be studied by dissolving it in a solvent and irradiating the solution with light of a wavelength corresponding to the excitation energy of the solute. No matter how dilute the solution, energy will only be absorbed by the solute under these circumstances, so that the specific reactions of the excited solute may be observed. On the other hand, if a solution is exposed to ionising radiations, energy will be absorbed by both the solute and solvent molecules, and excited species and ions arising from both solute and solvent will be produced. This gives rise to the concept of a *direct effect* of the radiation on the solute and an *indirect effect*, in which excited species and ions of the solvent subsequently react with the solute. It will be understood that if the solution is very dilute, any direct effect on the solute molecules will be negligible in comparison with the indirect effect. The reactions which will be observed will be those resulting from the interactions of excited species from the solvent with themselves, or with any solute which is present. In this way, fairly specific effects can be produced, so that with ionising radiations there is a non-uniform and non-specific energy absorption, resulting in specific effects.

Very often a field of study will develop from a particular method of initiation of chemical reactions. In this way, photochemistry is the study of reactions initiated largely by ultraviolet radiation. In a similar manner, the reactions initiated by ionising radiations are classed together as *radiation chemistry*. Radiation chemistry may thus be defined as the study of chemical reactions resulting from the absorption of energy from ionising radiations.

SYSTEMS OF UNITS

Before one considers the origin and development of radiation chemistry the units in which various relevant physical quantities are measured must be established. The fundamental dimensions in which mechanical quantities may be expressed are chosen as mass, length and time. If mass is measured in grammes, g, length in centimetres, cm, and time in seconds, s, the resulting units are CGS (centimetre-gramme-second) units. Alternatively, if mass is measured in kilo-grammes, kg, length in metres, m, and time in seconds, the resulting units are MKS (metre-kilogramme-second) units. In general, quanti-ties of interest to chemists have been expressed in the CGS system, but an international system of units is being introduced at the present time. This system, based on the MKS system, is known as the *Système International d'Unités* (usually abbreviated to SI). New publications will probably be gradually adopting SI units but there will, of course, still exist a great deal of literature based on the CGS system. For some time to come, then, a familiarity with both CGS units and SI units will be desirable. The CGS and SI units for some mechanical physical quantities are given in *Tables 1.1* and *1.2*.

When the CGS or MKS systems are extended to cover electrical quantities, a fourth quantity must be defined in addition to mass, length and time. Two choices have been in common use in the CGS

Table 1.1 CGS and SI units

Quantity	Symbol	CGS unit	SI unit
mass	m	g	kg
length	l	cm	m
time	t	s	s
velocity	u	cm s^{-1}	m s^{-1}
acceleration	a	cm s^{-2}	m s^{-2}

Table 1.2 CGS and SI units

Quantity	Symbol	Name of unit		Definition of unit	
		CGS	SI	CGS	SI
force	F	dyne	newton (N)	g cm s^{-2}	kg m s^{-2}
work	w	erg	joule (J)	g cm^2 s^{-2}	kg m^2 s^{-2}

From *Table 1.2* it will be appreciated that

$$1 \text{ N} = 10^5 \text{ dyne}$$
and
$$1 \text{ J} = 10^7 \text{ erg}$$

system but only one of them need be considered here. This is to choose electric current as the fourth quantity from which charge and other electrical quantities may be subsequently defined. This choice gives rise to electromagnetic units (e.m.u.). In practice, however, the e.m.u. of electrical quantities were not very convenient, as they tended to be rather larger or very much smaller than the amounts of these quantities encountered in practical situations. In the CGS system, then, a system of practical units was devised. These were the ampere, coulomb, volt and ohm, and were defined as multiples of powers of ten of e.m.u. Since these practical units were in everyday use, it was convenient to make them the basis of the SI electrical units. This was achieved by defining the SI ampere in such a way as to make it identical with the practical ampere. The SI units of electrical quantities and their definitions are given in *Table 1.3*.

Table 1.3 SI electrical units

Quantity	Name of unit	Symbol for unit	Definition of unit
current	ampere	A	
charge	coulomb	C	A s
power	watt	W	J s^{-1}
potential difference	volt	V	W A^{-1} (= J A^{-1} s^{-1})
resistance	ohm	Ω	V A^{-1}

In the SI there are seven basic physical quantities, four of which—mass, length, time and current—have been encountered above. The remaining three basic quantities are thermodynamic temperature, T, amount of substance, n, and luminous intensity, I_v. The last of these has no relevance to the present text and will be neglected. The SI unit of thermodynamic temperature is the kelvin (K) and the SI unit of amount of substance is the mole (mol). The mole is defined as 'the amount of substance which contains as many elementary units as there are atoms in 0·012 kg of carbon-12. The elementary unit must be specified and may be an atom, a molecule, an ion, a radical, an electron, a photon, etc., or a specified group of such entities.' In 0·012 kg of carbon-12 there are approximately $6·02 \times 10^{23}$ atoms, so that within the limit of this approximation a mole of e^- means $6·02 \times 10^{23}$ electrons. A mole of H_2SO_4 means $6·02 \times 10^{23}$ particles of H_2SO_4 and a mole of $\frac{1}{2}H_2SO_4$ means $6·02 \times 10^{23}$ particles of $\frac{1}{2}H_2SO_4$. It will be appreciated that the term 'a mole of sulphuric acid' is somewhat ambiguous unless the elementary unit is specified.

All quantities in the SI, other than the basic quantities, are derived from the basic quantities by definitions involving only multiplication, division, differentiation or integration. *Table 1.4* gives some examples of SI nomenclature for some derived physical quantities.

Table 1.4 Derived physical quantities

Quantity	Name of SI unit	Symbol for SI unit
area	square metre	m^2
volume	cubic metre	m^3
density	kilogramme per cubic metre	$kg\ m^{-3}$
pressure	newton per square metre	$N\ m^{-2}$
energy	joule	$J\ (= N\ m)$
concentration	mole per cubic metre	$mol\ m^{-3}$
molality	mole per kilogramme	$mol\ kg^{-1}$

A wide range of magnitudes is encountered in physical science, and in order to avoid writing very large and very small numbers, the SI has adopted certain prefixes to indicate decimal fractions or multiples of units. These prefixes are given in *Table 1.5*.

Table 1.5 Prefixes for SI units

Fraction	Prefix	Symbol	Multiple	Prefix	Symbol
10^{-1}	deci	d	10	deka	da
10^{-2}	centi	c	10^2	hecto	h
10^{-3}	milli	m	10^3	kilo	k
10^{-6}	micro	μ	10^6	mega	M
10^{-9}	nano	n	10^9	giga	G
10^{-12}	pico	p	10^{12}	tera	T
10^{-15}	femto	f			
10^{-18}	atto	a			

It should be noted that in the SI the only permitted units of energy are the joule, and decimal fractions and multiples of the joule. In radiation chemistry the commonly used units of energy have been the erg and the electronvolt, particularly the latter, which is the amount of energy possessed by an electron when it has been accelerated by a potential difference of one volt. As many results in radiation chemistry are expressed in terms of ergs and electronvolts, it will be useful to have some idea of the equivalent energy in SI units. The energy of an electron accelerated by one volt must be equal to the work done when the charge carried by an electron ($1\cdot6 \times 10^{-19}$ C) is transferred through a potential difference of one volt, which is given by

$$w = 1\cdot6 \times 10^{-19} \text{ C} \times 1 \text{ V}$$
$$= 1\cdot6 \times 10^{-19} \text{ J}$$

Thus

$$1 \text{ eV} = 1\cdot6 \times 10^{-19} \text{ J}$$

or

$$1 \text{ eV} = 0\cdot16 \text{ aJ}$$

Because of the existing method of presentation of the results of radiation chemistry, many of the results quoted in this book will have to be in non-SI units, but future conversion to SI units should be readily accomplished on the basis of the above considerations.

Another practice which will be followed concerns the headings of tables of figures and the labelling of the axes of graphs. In stating the value of a physical quantity, the following relationship applies:

$$\text{physical quantity} = \text{number} \times \text{unit}$$

For example,

$$p = 2\cdot3 \text{ N m}^{-2}$$

This equation could equally well be written

$$p/\text{N m}^{-2} = 2\cdot3$$

A further example is provided by

$$w = 1\cdot6 \times 10^7 \text{ J}$$

which could be written as

$$w/10^7 \text{ J} = 1\cdot6$$

In the presentation of tables of values of physical quantities where only pure numbers appear in a particular column, the heading of the column will be written as the symbol for the physical quantity *divided* by the units in which the quantity is expressed. Similar considerations will apply to the labelling of the axes of graphs.

THE ORIGIN OF RADIATION CHEMISTRY

The field of study known as radiation chemistry can be said to stem from Röntgen's discovery of X-rays in 1895. Röntgen was carrying out experiments on the discharge of electricity through gases, in the hope of obtaining information on the nature of cathode rays, when he noticed that a sheet of paper coated with barium platinocyanide fluoresced when the discharge was operating. The fluorescence was observed even when the discharge tube was enclosed in black paper which absorbed any ultraviolet light from the discharge; the fluorescence was thus due to some hitherto unknown type of radiation, which was labelled X-radiation.

In a series of experiments Röntgen was able to elucidate some of the properties of X-rays. He found that they were very penetrating and could pass through substances which were opaque to ordinary light. All substances reduce the intensity of the rays, and the absorbing power of a material depends to some extent on its density and to some extent on its relative atomic mass. Light substances, and those containing low relative atomic mass materials, are comparatively transparent, and Röntgen showed that the X-rays had the power to blacken a photographic plate wrapped in black paper. He also showed that a gold leaf electroscope could be discharged when exposed to X-radiation. Thomson also noticed this phenomenon, and the cause was soon traced to the ionisation of the air molecules in the electroscope. It was thus shown that X-rays were capable of causing ionisation in gases.

Subsequent work has shown that X-rays are electromagnetic radiation, similar to light but of much shorter wavelength (< 5 nm), which are produced by the bombardment of matter by electrons.

Röntgen had shown that X-rays could produce fluorescence in several substances and he had also reported that the X-rays emanated from that part of the discharge tube opposite the cathode, and that the inner wall of the tube in this region fluoresced. It was this connection between X-rays and fluorescence which attracted the attention of Becquerel in 1896. Becquerel had been interested in the phenomenon of fluorescence for some time, and several years earlier he had prepared potassium uranyl sulphate and noticed its pronounced fluorescence under the influence of ultraviolet light. He wondered whether fluorescence was accompanied by the emission of X-rays and, in order to test this, he exposed several fluorescent materials to sunlight and placed them on a photographic plate wrapped in black paper. If X-ray emission occurred, the photographic plate would be blackened. These experiments were unsuccessful until Becquerel used the potassium uranyl sulphate salt. With this salt the photographic plate was blackened, and it was shown that interposing a metal plate reduced the intensity of the blackening, so that the radiation was more strongly absorbed by metal than by paper. Becquerel also found that the radiations from the salt would discharge an electroscope, so that their properties seemed similar to the known properties of X-rays. It was noticed, however, that the effects were independent of the intensity of the light producing the fluorescence and that the radiations were generated by crystals of the salt which had been prepared and kept in darkness. Further investigation revealed that the radiation was emitted by other uranium salts, including those which did not fluoresce, and the intensity of the radiation, as measured by the blackening of a photographic plate, was proportional to the quantity of uranium in the sample.

In 1898 Pierre and Marie Curie showed that the radiation emitted by uranium compounds was an atomic phenomenon which did not depend on the chemical or physical state of the active element. The phenomenon was termed *radioactivity* by the Curies. During the year 1898 a great deal of work was done, mostly by the Curies, which showed that thorium compounds were also radioactive and which led to the discovery of two new radioactive elements, polonium and radium, both of which were intensely radioactive. Marie Curie used an electrical method of detecting the radiations from the radioactive substances rather than a photographic plate technique. Thomson had already pointed out that the ability of X-rays to form ions would provide a more sensitive method of detection of X-rays and Marie Curie used a similar method for the detection of radiations from radioactive substances, based on their ionising power. The radioactive substance was placed between two parallel plates, across which a voltage was applied. The radiation ionised the gas molecules in the space between the plates, and the negative and positive ions travelled to the positively and negatively charged plates, respectively. The migration of the ions to the plates constituted a current, which was measured with an electrometer. This device was the forerunner of the modern ionisation chamber. The amount of ionisation produced by the radiation was proportional to the activity of the sample and, hence, the current was a measure of the activity.

In 1899 Rutherford used a similar apparatus to study the nature of the radiation. By interposing metal foils between the source of radiation and the detector, he was able to show that there were two types of radiation. Whereas one of these was absorbed completely by a few thousandths of a centimetre of aluminium foil, the other required about a hundred times this thickness for absorption. The first type

B

was named α-radiation and the second type β-radiation. At about the same time, Villard showed that a third type of radiation, much more penetrating than β-radiation, could be obtained from some radio-active elements. This was called γ-radiation.

The nature of the three types of radiation provided by radioactive elements was largely resolved by the effects of magnetic and electro-static fields on the rays. α-rays were shown to consist of a stream of helium nuclei moving with velocities of about one-tenth the speed of light. β-rays were identified as electrons with velocities up to values of just less than the speed of light, and γ-rays were shown to be electromagnetic waves of the character of X-rays but usually with rather higher energies.

The α-rays emitted by a particular radioactive source are mono-energetic or have a few discrete energies. Monoenergetic α-rays can be characterised by a fairly well-defined range in a given absorber, which means that a beam of α-rays maintains the original number of particles in it until a certain distance has been traversed beyond which no particles are transmitted. β-rays, on the other hand, have a continuous spectrum of energies ranging up to a maximum value. As a result, not all β-rays from the same source penetrate the same distance into an absorber but penetrate to different extents, depend-ing on the energy of a particular β-particle. For β-rays, then, only a maximum range can be quoted.

The γ-rays emitted by radioactive sources are similar to α-rays in being either monoenergetic or having a small number of discrete energies, but unlike α-rays they do not have a definite range in matter. This difference is due to the different mechanisms by which α-rays and γ-rays lose their energy; this matter will be dealt with in greater detail in Chapter 3. For the time being it is sufficient to note that the intensity of a beam of γ-rays decreases logarithmically as it passes through matter according to the relation

$$I = I_0 e^{-\mu x} \qquad (1.1)$$

where I_0 is the initial intensity of the beam, I is the intensity after the beam has travelled a distance x into the absorber, and μ is a constant for a given absorber and a particular γ-ray energy. Phe-nomena which follow logarithmic laws can be characterised by a *half-value* and this is the case for γ-rays.

Equation (1.1) may be written:

$$\ln (I/I_0) = - \mu x$$

or

$$2 \cdot 303 \log (I/I_0) = - \mu x$$

If $x_{\frac{1}{2}}$ is the thickness of absorber required to reduce the intensity of the beam to half the original value, then

$$2{\cdot}303 \log (\tfrac{1}{2}) = -\mu x_{\frac{1}{2}}$$

or

$$x_{\frac{1}{2}} = \frac{2{\cdot}303 \log 2}{\mu} = \frac{0{\cdot}693}{\mu} \tag{1.2}$$

As μ is a constant for a certain γ-ray energy and a given absorber, $x_{\frac{1}{2}}$, known as the *half-value thickness*, will also be a constant under these conditions. Thus, although γ-rays do not have a definite range, they can be characterised by a half-value thickness.

Soon after the discovery of radioactivity, the chemical effects of the radiations from radioactive substances aroused interest. Hydrated radium salts evolved gas continuously and solutions of radium bromide were found to evolve a mixture of oxygen and hydrogen. Attention was devoted largely to the reactions of gases initiated by α-particles, as these systems were more amenable to study with the experimental techniques then available. The main source of radiation was radium, which was used chiefly to provide α-particles. Some work was done with liquids, but the range of α-particles in liquids is very small and even when the more penetrating β- and γ-radiations were used, the sources were small and long periods of irradiation were necessary to yield detectable amounts of products.

The advent of X-ray machines which produced radiation of greater intensity permitted the investigation of liquid and solid systems. At the same time, the biological effects of X-rays directed attention to the radiation chemistry of aqueous systems so that, up to about 1940, radiation chemistry expanded from the study of gaseous systems, largely with α-particles, to the study of aqueous systems, largely with X-rays. The main development of radiation chemistry came with the establishment of atomic energy programmes during World War II. These programmes gave rise to machines which could accelerate particles such as electrons, protons and other heavy positive ions to very high energies and provided sources of high-energy radiations. In addition, the production of artificial radioactive elements, such as cobalt-60, provided intense sources of γ-radiation.

The current sources of high-energy radiations are discussed in the next chapter, but it might be useful to clarify some points of terminology at this stage. High-energy helium nuclei, electrons and electromagnetic radiation can be obtained from machines or from radioactive sources. The terms α-particles, β-particles and γ-rays will be retained for the radiations emanating from radioactive sources, to distinguish them from the artificially produced radiations, which will be called helium nuclei, electrons and X-rays. β-particle electrons may be denoted by the symbol β^-, and electrons from machines and those resulting from ionisations in a medium will be denoted by the symbol e^-. Similarly, positrons from nuclei may be represented by β^+, while other positrons may be represented by e^+.

2
Sources of ionising radiations

There are two main types of source from which ionising radiations may be obtained. First, radiations may be obtained from radioactive materials, either natural or artificial; and second, they may be obtained from machines which provide high-speed electrons or ions. To the system which is being irradiated, the source of the radiations is immaterial. The over-all effect of exposing any material to ionising radiations is the transfer of energy to the system, resulting in the ionisation and excitation of some of its constituent atoms and molecules. This in turn may lead to chemical reaction, the nature of which will be quite independent of the source. One exception to this is the type of reaction in which the source is located internally in the absorbing medium when the radiation chemistry of the source itself is being examined. From these considerations it will be understood that a radiation source can be internal or external to the medium under investigation. The great majority of experimental arrangements make use of external sources, and these will be discussed first.

EXTERNAL SOURCES

RADIOACTIVE ISOTOPES

The chemical reactions produced in a medium subjected to ionising radiations may depend on the density per unit length of track of the ionisation and excitation events. This in turn is characteristic of the

13

Table 2.1 Radiations from radioactive isotopes used as external sources

Isotope	Origin of isotope	Half-life	Important radiation	Average energy	Penetration/m	
					air	water
Cobalt-60	^{59}Co(n,γ)^{60}Co	5·27 years	γ	1·25 MeV (0·200 pJ)		1·1 × 10^{-1}
Caesium-137	Fission product	30 years	γ	0·66 MeV (0·106 pJ)		8·1 × 10^{-2}
						HVT
Strontium/Yttrium-90	Fission product	28 years	β	2·25 MeV (0·36 pJ)	1·85	1·8 × 10^{-3}
Polonium-210	Natural and from ^{209}Bi(n,γ)^{210}Bi–^{210}Po	138 days	α	5·30 MeV (0·85 pJ)	3·8 × 10^{-2}	3·84 × 10^{-5}
						Range

type of radiation employed. It is important, therefore, to know some of the properties of the different ionising radiations which can be used to initiate chemical reactions. *Table 2.1* gives these properties of the important radiations from a few of the radioactive isotopes which have been commonly used as external sources.

Before discussing these sources in detail, it will be useful to define two units. When a radioactive source emits a particle, it is said to undergo a disintegration, and the *activity* of a source is measured by the number of disintegrations occurring per unit time. The unit of activity commonly used is the *curie,* Ci, which is equal to $3 \cdot 7 \times 10^{10}$ disintegrations per second. Specific activity is the activity per unit mass. All radioactive sources decay as the disintegration occurs and a measure of the rate of decay is given by the *half-life*, which is the time taken for the activity to decrease by one-half. The shorter the half-life of an isotope used as a source of radiation, the sooner it will have to be replaced; this should be borne in mind when one constructs radioactive sources.

The other unit requiring definition refers to the energy absorbed by a medium exposed to ionising radiations. This is usually expressed in *rads,* where one rad is equal to 100 ergs per gramme of absorber $(10^{-2} \text{ J kg}^{-1})$. The energy absorbed by a medium is frequently referred to as the dose which it has received, and the rate of energy absorption is known as the dose rate.

External γ-sources

It has been pointed out in Chapter 1 that γ-rays are the most penetrating of the radiations from radioactive sources and that they are never completely absorbed. This comparison is made quantitatively in *Table 2.1*. The absorption of γ-rays in air can be neglected and the walls of the radioisotope container should reduce the intensity of the rays only a small fraction. Isotopes which emit γ-rays are thus ideal as external sources. The production of a significant amount of chemical change in radiation chemistry requires the absorption of large doses (> 500 rad (5 J kg^{-1})) and, therefore, rather large sources. These can be conveniently obtained for γ-rays from the nuclear reaction of cobalt-59 with thermal (slow) neutrons. This reaction produces radioactive cobalt-60 and may be written:

$$^{59}\text{Co} + {}^{1}\text{n} + {}^{60}\text{Co} + \gamma$$

Nuclear reactions are frequently written in an abbreviated form by which the above reaction is represented by:

$$^{59}\text{Co}(n,\gamma)^{60}\text{Co}$$

The reaction is carried out in a nuclear reactor and the cobalt is irradiated in the form of metal rods which are sealed in stainless steel cylinders. With a thermal neutron flux of 10^{16} neutrons m^{-2} s^{-1}, the cobalt will attain a specific activity of 1 Ci g^{-1} after 1 year's exposure, and after 20 years' exposure the specific activity will be 10 Ci g^{-1}. The radioactive cobalt-60 decays by emission of a β-particle to an excited nickel-60 nucleus, which immediately achieves stability by emitting two γ-rays. These processes may be represented by the following decay scheme:

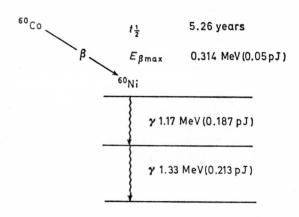

In the case of a radioactive cobalt-60 source, no chemical work need be done on the rods once the cobalt-60 has been formed in the reactor. This greatly simplifies any subsequent handling problem. A number of these activated rods can be assembled by remote handling techniques to give sources of extremely high activity. Sources emitting 7.4×10^{14} photons per second have been successfully assembled and used for radiation purposes. Naturally, such large quantities of radioactivity pose a number of health hazards. There should be no chance of accidental ingestion of the radioactive isotope, as the cobalt is in the form of the metal, which is completely sealed in stainless steel. There will, however, be a very serious hazard of external exposure to radiation. Five rad (5×10^{-2} J kg^{-1}) is the maximum dose which anyone working with ionising radiation is permitted to receive in 1 year (maximum permissible level recommended by the International Commission for Radiological Protection). As this is the dose which would be received at a distance of 1 ft from a 0·3 Ci cobalt-60 point source, it will be readily understood that massive shielding and elaborate safeguards are required for sources which have activities greater than 100 Ci. Two main types of shielded cobalt-60 radiation units have been designed, which allow very large quantities of the radioactive isotope to be used safely These may be called *the well-type source* and *the labyrinth source*.

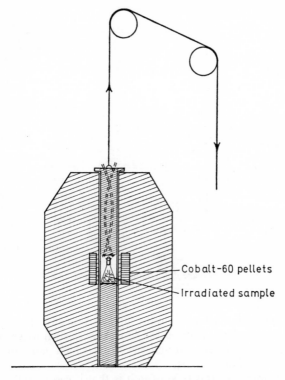

Cobalt-60 pellets

Irradiated sample

Figure 2.1. Well-type cobalt-60 source

In the well-type source the cobalt-60 rods are arranged in a cylindrical fashion around a cavity in which irradiation is carried out (*Figure 2.1*). The whole unit is placed in a large barrel of lead which reduces the radiation outside the shield to not more than the maximum permissible level (*c.* 2·5 mrad h^{-1} (7 pJ kg^{-1} s^{-1})). In another similar unit the cobalt source is sunk into the ground with sufficient concrete or lead shielding on top to prevent any upward radiation hazard. The actual irradiation chamber is part of a plug which can be moved up and down in the middle of the cobalt-60 rods. Above and below the irradiation chamber, the plug carries shielding material to prevent any radiation scattered vertically from reaching the top of the unit, whether the chamber is in the 'up' or 'down' (irradiation) position. In different units the plug can be moved manually, mechanically or by compressed air. Approximately 2 dm³ of experimental volume can be irradiated with this type of unit. The radiation vessel is

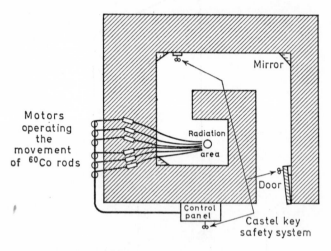

Figure 2.2.   Labyrinth source

inserted in the plug while it is in the 'up' position and the plug is then lowered for irradiation purposes. Any services such as gas or electricity can be supplied through the ducts which run through the shielded portion of the plug.

The advantage of this type of unit is that it can be set up in any laboratory. On the other hand, there is the disadvantage that the radiation volume is limited to about 2 dm³.

Where a large amount of space around the radiation source is required, the labyrinth source (*Figure 2.2*) has all the advantages. The radiation area is located in a concrete bunker or cellar which has a labyrinth access to prevent the escape of any stray radiation. The cobalt-60 rods can be moved in tubes by means of cables from a 'safe' position inside the concrete walls to an irradiation position around a radiation vessel in the radiation area. The whole arrangement is safeguarded by means of warning lights, sounds and an interlocking system. The latter (Castel key system) ensures that an operator cannot enter the radiation area without having first moved the cobalt rods to the safe position. The one key, which controls the movement of the cobalt rods and which also opens the access door, must be taken into the radiation area to switch on the light and prevent the continuous sounding of an alarm. Any experiment can then be set up in the middle of, or around, the tubes which carry the cobalt rods into the radiation area, which can be provided with any type of service. When the operator leaves the chamber, the key is removed and must be used within 30 s to lock the access door so that the cobalt-60 rods may be moved along the tubes into the irradiation position. There are additional safety buttons inside the chamber by which the movement of the cobalt-60 rods can be instantly stopped.

The shielding around the safe position of the cobalt-60 rods is so arranged that an operator in the radiation area receives no dose above the maximum permissible level. It should also not be possible to receive a dose above this level outside the labyrinth irrespective of whether the source is in the safe or irradiation position. The storage or safe position of the cobalt-60 rods can be horizontal, as shown in *Figure 2.2*, or vertical—whichever is the most convenient.

The advantages of this type of radiation chamber lie in the very large volumes which can be irradiated simultaneously and the ease with which any other services can be used. In some units the ends of the tubes in the radiation area are flexible, so that variations in source geometry and dose rate can be achieved. The disadvantage of the labyrinth source is the large amount of space which is required and the permanent building work which must be undertaken.

The labyrinth and well-type sources have been combined in industrial irradiation systems where the products are moved continuously past the cobalt-60 source. Here the source rods are stacked in a vertical frame to provide enormous activities. When not in use, the source is lowered into a 'safe' well. In the irradiation position the source is located in a large labyrinth concrete bunker. The product to be irradiated is fed on a continuous belt into and out of the radiation area, and it can be so arranged that the material is moved past the source more than once.

External β-sources

Where an external β-source is required, strontium/yttrium-90 appears to be the most suitable isotopic source. It is a product of nuclear fission reactions, and is separated from other fission products and bonded into strips or on to plates of silver. In an alternative form the strontium is made into a ceramic or sintered with titanium metal. In the latter form it is sealed into stainless steel capsules.

As β-radiation is readily absorbed in air (see *Table 2.1*), the material to be irradiated must be placed as near as possible to the source. Great care must be taken not to damage the surface of the source in an attempt to obtain a high dose rate. A broken surface may lead to the escape of the radioactive material, which might be ingested by the operator. This would give rise to an internal radiation hazard, as strontium-90 accumulates in bone.

Although β-sources are more easily shielded than γ-sources, they still present an external radiation hazard. Near the source the intensity of β-particles will be high (up to $3 \cdot 7 \times 10^{14}$ β-particles m^{-2} s^{-1} can be obtained from these sources) and eyes are particularly sensitive to intense β-radiation. Furthermore, if the energy of the β-particles is high, as it is in the case of $^{90}Sr/^{90}Y$, X-rays can be generated and constitute an additional hazard. The production of X-rays can occur whenever high-speed electrons are decelerated, and the efficiency of production increases as the atomic number of the stopping material increases. Thus, in the shielding of β-sources, material of low atomic number is placed next to the source, in order to minimise the production of X-rays. This point will be emphasised more fully in the next chapter.

β-sources are of little use for the irradiation of bulk material, as the penetration of β-rays is so small. Rather, they provide a very intense densely ionising source of radiation, which can be used successfully on thin films, or they may be used to discharge static electricity.

External α-sources

α-particles from radioactive isotopes show even less penetration than β-particles, and thus it is very seldom that α-emitters are employed as external sources. Polonium-210 deposited on a platinum disc can be bought as a source from the Radiochemical Centre at Amersham. The source is covered with a thin mica window, but despite this an internal health hazard may develop. First, the polonium 'walks' as a result of recoil reactions; and second, the high ionisation intensity of the α-particles causes the thin mica window to crack. The polonium-210 thus escapes from the source, which should therefore only be used inside a glove box.

Nuclear reactors and spent reactor fuel

Two sources for external radiation not mentioned in *Table 2.1* are the nuclear reactor and its spent nuclear fuel. Massive doses of radiation can be obtained inside nuclear reactors but the radiations are mixed in type and energy. Among these are slow and fast neutrons, which will induce radioactivity in the samples being irradiated and thus complicate their subsequent handling and analysis. Another difficulty is that the reactor may be running at a high temperature. Usually the radiations from a nuclear reactor are only used when some new constituent material for reactor construction is being investigated. Doses from nuclear reactors are sometimes given in pile units where 1 pile unit = 1 megarad (10^4 J kg^{-1}).

Fuel elements taken out of a reactor can be used as a source of γ-radiation. The 'hot' fuel rods are loaded into a frame and submerged in a cooling tank of water. The sample to be irradiated is submerged alongside. The arrangement is essentially similar to the well-type cobalt-60 sources. By the time the fuel elements are in the cooling pond, there should only be a few neutrons left. The α- and β-particles will be absorbed by the fuel cladding and the water. Thus a mixture of γ-rays will be available for irradiation. Dose rates of up to 8 Mrad h^{-1} (22 J kg^{-1} s^{-1}) can be obtained from such an arrangement.

MACHINE SOURCES

Radioactive isotopes are not suitable as external sources for α- and β-particles when the medium to be irradiated is a bulk liquid or solid. The penetration of these particles in condensed phases is very small, as they are emitted with relatively low energies. With accelerating machines, however, these types of particles can be given so much energy that they will penetrate quite deeply into dense media. To obtain a reasonable yield of products in radiation chemistry, high energies alone are not sufficient. It should be remembered that the intensity of the radiation must be high. The principle of all accelerating machines is the same, in so far as they accelerate charged particles to high energies by the application of electric fields. It has been mentioned above in connection with the shielding of radioactive β-particle sources that high-speed electrons can produce X-rays when they are decelerated by an absorber. It will be appreciated, then, that in addition to producing high-intensity beams of high-energy charged particles those accelerators which produce electron beams can be adapted to provide X-rays by allowing the electrons to impinge on a target. X-rays of lower energies are usually produced by the standard X-ray unit.

X-ray machines

The X-ray tube (*Figure 2.3*) of the standard X-ray unit is a gas discharge tube consisting of a highly evacuated glass envelope containing a heated filament cathode which supplies the electrons. The anode is the target on which the electrons impinge and is usually fixed and made of tungsten. The potential difference applied across X-ray tubes used in medical radiography and X-ray crystallography is about 60 kV, and that employed in industrial radiography is about

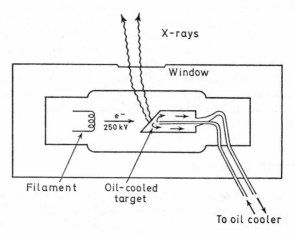

Figure 2.3.   X-ray tube

150 kV. For radiation chemical purposes and deep X-ray therapy voltages up to 250 kV are used. The electrons from the cathode are accelerated through these voltages and are absorbed by the target anode. Whenever high-speed electrons are decelerated and stopped in this way, some of their kinetic energy is converted into X-rays. This occurs by two processes.

First, the electrons may decelerate in the region of an atomic nucleus, giving rise to X-rays with a continuous energy spectrum from the maximum energy of the incident electrons down to almost zero. This radiation is known as 'bremsstrahlung', which simply means braking radiation. The greater the atomic number of the absorber, the greater is the proportion of the electron's energy which is converted to bremsstrahlung and this is one reason for choosing tungsten as the target.

In another process the incident electrons can ionise the target atoms by ejecting an atomic electron from an inner shell. As the electrons from the outer shells drop into the vacancy so created and thus achieve a lower energy state, the energy which they lose is emitted as X-radiation. These X-rays, however, have well-defined energies corresponding to the difference in energy of the electron which falls from an outer shell to the inner shell. Such X-rays of well-defined wavelength are known as the 'characteristic X-rays' of the target element. Thus the sum total of X-radiation produced consists of the bremsstrahlung X-rays, which cover a wide range of energy up to the applied voltage, and, superimposed on these, the characteristic X-rays of the target element. For radiation chemical purposes only the bremsstrahlung need be considered, and the highest intensity of the bremsstrahlung corresponds to an energy of about one-third of the maximum energy. Metal filters can be used to narrow the wide energy range of the X-ray beam, but this practice leads to a simultaneous reduction of intensity, which will decrease the rate of any radiation chemical reaction.

Most of the early research workers used the X-ray tube as their radiation source because it was the most powerful one available at that time. In most cases where X-rays are used in radiation chemistry, high-voltage tubes are used without any filters. For most experimental purposes this is quite satisfactory; but where the interactions and radiation yields are dependent upon energy, corrections have to be made. It is therefore essential that the conditions of operation (i.e. voltage, current, filters) of an X-ray tube are accurately stated.

Van de Graaff machines

The Van de Graaff generator produces a high voltage which can be used in conjunction with an *accelerating tube* to give a continuous stream of high-energy electrons, or by reversal of the potential difference along the accelerating tube, positive ions. The high voltage is produced by having a continuous moving belt which deposits electrostatic charge on to a hollow spherical electrode. In this way the charge builds up and the maximum potentials which can be obtained (5–30 MV) are limited only by insulation difficulties. The high-voltage electrode is thus surrounded by gas at high pressure to minimise discharge to the surroundings.

The accelerating tube itself is essentially a system of accelerating electrodes across which the high voltage from the Van de Graaff generator is applied. The entire arrangement is illustrated in *Figure 2.4*. The tube is a long cylinder of metal divided into sections. The potential of the generator is dropped uniformly over the whole length of the tube, so that there are equal differences of potential between

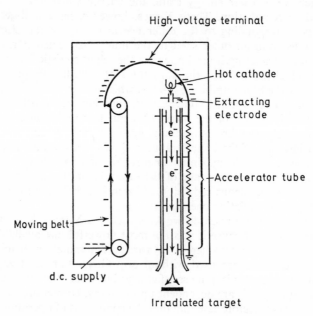

Figure 2.4. Van de Graaff generator and accelerating tube

each successive section. The source of charged particles is located near the end of the tube, which is connected directly to the electrode of the Van de Graaff generator. The metal sections of the accelerating tube are surrounded by an insulating tube of glass or porcelain, the inside being evacuated to facilitate the passage of the charged particles and the outside being surrounded by high-pressure gas to minimise electrical leakage. If positive ions are to be accelerated, the generator electrode will carry a high positive potential, so that the ions travelling down the tube are accelerated by the successive potential differences across the gaps between the sections of the tube. If electrons are to be accelerated, the polarity of the generator electrode will be reversed. The gaps in the tube also help to focus the beam of particles.

The linear accelerator

In the Van de Graaff machine the full potential corresponding to the final energy of the charged particles must be provided and the operation is thus limited by insulation problems. These problems are overcome by linear accelerators in which 50–100 keV electrons are injected into an evacuated tube and accelerated by riding on radio-frequency (electromagnetic) waves, which are produced by klystron valves and propagated down the tube, which acts as a waveguide. Circular irises are inserted into the tube at critical points to allow only those wave phases to pass which accelerate the electrons. Electron energies above 600 MeV (96 pJ) have been achieved by this means and the electrons so produced can be used as primary radiation particles. A special adaption of this machine is used in pulse radiolysis (see Chapter 4), where radiation doses of several thousand rads are required within a microsecond.

For X-ray production the electrons from a linear accelerator are fired into a thin piece of high-atomic-number material from which the electromagnetic radiation emerges on the far side. At these high energies most of the X-ray emission will be in the forward direction.

The betatron

The betatron can be thought of as a transformer in which the secondary winding has been replaced by an evacuated circular tube concentric with the primary winding. Electrons are injected in pulses into this tube, known as 'the doughnut', and the magnetic and electric fields created by the alternating current in the primary constrain the electrons to a circular path within the doughnut and accelerate them. The electrons are injected into the doughnut with an electron gun and are extracted by means of an electric field between condenser plates. Targets such as tungsten wire for X-ray production can then be placed in the path of the electrons.

The cyclotron

In the cyclotron, multiple accelerations by high-frequency potentials are employed as in the linear accelerator. In the cyclotron, however, the path of the charged particles is spiral rather than linear. The particles travel round in a hollow disc which is split diagonally into

c

two separate electrodes known as 'dees'. A uniform magnetic field is applied across the disc which would normally constrain a charged particle to a circular path within the dees. As the particle is accelerated, every time it crosses the gap between the dees its orbit increases, so that it spirals outwards. This machine has been improved for high-energy particles, where a relativistic increase in the mass of the particle requires the frequency of the applied voltage to decrease as the energy of the particle increases. In this modification the machine is known as a frequency-modulated cyclotron or a *synchrocyclotron*.

The synchrotron

The synchrotron may be regarded, in principle, as a combination of a cyclotron and a betatron. The paths of the charged particles are defined by the a.c. electromagnet of the betatron and the acceleration is provided by a high-frequency voltage, as in the cyclotron. This machine is used largely to study nuclear interactions.

From all these orbiting machines, particles with energies less than the maximum can be produced by extracting them before they have completed the full number of orbits. *Table 2.2* summarises some types of particle accelerators available, showing the type of particle accelerated and giving maximum energies.

Apart from the X-ray unit, all the machine sources discussed above are expensive, require expert operators and are thus not readily available. It follows that the external sources in greatest use at present are the cobalt-60 source and the deep-therapy X-ray tube.

Table 2.2 Particle accelerators

Machine	Accelerated particle	Maximum energy
Van de Graaff	positive ion	30 MeV
	electrons	5 MeV
Linear accelerator	positive ions	
	electrons	600 MeV
Cyclotron	positive ions	
Synchrocyclotron	positive ions	
Cockroft–Walton accelerator	positive ions	
Betatron	electrons	300 MeV
Synchrotron	positive ions	1·5 GeV
	electrons	

INTERNAL SOURCES

An internal source is one which is located in the medium under investigation. It can be heterogeneously or homogeneously distributed within the medium. A heterogeneous internal source not only supplies the penetrating radiations but may act as an impurity, its surface acting as a catalyst and complicating the radiation-induced chemical reactions. Homogeneous sources are less likely to act as catalysts but may also interfere. As a result, internal sources are seldom deliberately chosen by radiation chemists. Most radiation chemical work with internal sources therefore arises where the chemistry of the source itself is under investigation. Studies of the chemistry of polonium-210 and the transuranic elements have produced most of the work on the effects of internal α-sources and most of the information on internal β-sources has come from research with carbon-14 and tritium-labelled organic compounds.

In addition to chemical reactions arising from purely ionising radiations, there are many which occur as a result of the nuclear changes undergone by the internal source. 'Szilard–Chalmers' reactions (see Chapter 3) are often the result of nuclear recoil processes. On the whole, the changes produced by internal sources are the same as those brought about by external sources and differ only in extent.

RADIATION VESSELS

For studying the chemical effects of ionising radiations any vessel will suffice, provided that the walls absorb a minimum of the radiations and do not interfere with the radiation products and their subsequent chemical reactions. Very often the vessels are made of glass, which discolours on prolonged irradiation. The vessel is usually placed as near as possible to the radiation source so that the maximum dose can be absorbed in a given time. Where the irradiation has to be carried out in the absence of air or in the presence of another gas, special vessels with the appropriate taps must be used.

The observation of physical properties during irradiation requires remote control of the system. Thus, in pulse radiolysis experiments where optical absorption has to be measured, the beam of light is passed through the medium at right angles to the radiation pulse. The vessel, in this case, is made of quartz and the beam which has passed through the solution is reflected by means of a series of mirrors to the outside of the radiation area. Here it is analysed for spectral absorption due to the transient species occurring, which may only have lifetimes of microseconds. The disappearance of the transients can be followed by means of a high-speed oscilloscope, which can measure changes in absorbance with a photomultiplier against a microsecond or even nanosecond time base.

Other properties which have been observed during irradiation include electrical conductance, paramagnetic resonance, gas formation, membrane phenomena, catalysis, polymer creep and tensile strength. In all cases special arrangements must be made to see that the radiations will not interfere with the detectors and that the results can be recorded or measured outside the radiation area.

GLOW DISCHARGE ELECTROLYSIS

The technique of glow discharge electrolysis could be classed as a machine source of ionising radiation, but it is considered separately here as it is rather a specialised technique, which at first glance may seem rather remote from radiation chemistry. In a conventional d.c. gas discharge between two metal electrodes in a gas at low pressure, the potential gradient between the electrodes is not uniform but has the form illustrated in *Figure 2.5*. There is a small potential drop near

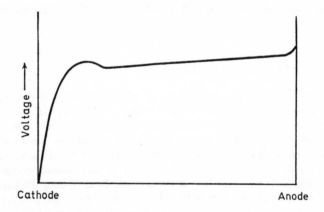

Figure 2.5.   Electric field in a d.c. glow discharge

the anode, after which the potential changes very little until the neighbourhood of the cathode is reached. It can be seen from *Figure 2.5* that most of the potential drop across the discharge occurs over a short distance near the cathode. This is called 'the cathode fall'. Positive ions formed in the discharge travel towards the cathode and are accelerated by the intense electric field of the cathode fall, so that they impinge upon the cathode with relatively high energies.

In glow discharge electrolysis a glow discharge is struck, not between two metal electrodes, but between a metal electrode and the surface of a solution. In most cases the metal electrode is the anode and the solution–gas interface is the cathode. The gas above the

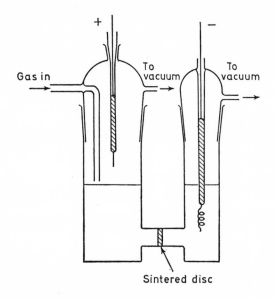

Figure 2.6.   Glow discharge electrolysis cell

solution can be chosen at will and is pumped down to a pressure of about 50 Torr (*c.* 6·7 kN m^{-2}), the minimum pressure attainable being limited by the vapour pressure of the solution at the temperature of the experiment. Positive ions travelling towards the liquid surface are accelerated by the cathode fall and enter the solution with a distribution of energies covering the range up to the accelerating potential. In the usual practice the cathode fall is about 400 V and the average energy of the ions entering the solution will be about 100 eV (16 aJ). The discharge itself is in the form of a cone with the apex at the metal electrode and the base on the solution surface. A typical vessel for glow discharge electrolysis is shown in *Figure 2.6*.

It will be appreciated that an ionising particle with an energy of 100 eV(16 aJ) has a very low energy in the context of radiation chemistry. These ions penetrate only a very short distance into the solution and give rise to a region of high concentration of active species just below the surface of the solution over a small area of the surface. Although the energy of the ions is low, very high dose rates can be achieved at moderate discharge currents (c. 75 mA). The technique of glow discharge electrolysis is thus an extension of conventional radiation chemistry to very low energies and very high dose rates.

3
The behaviour of ionising radiations

The chemical changes studied in radiation chemistry are the eventual consequences of the absorption by matter of the energy of ionising radiations. The mechanisms by which the radiation transfers energy to matter through which it passes depend very largely upon the type of radiation. From this point of view ionising radiations can be divided into three classes:

(1) charged particles (electrons and heavy positive ions);
(2) neutral particles (neutrons);
(3) electromagnetic radiation (X-rays and γ-rays).

Before we deal with these three classes in detail, it will probably be useful to make some general observations on the behaviour of particles. A fast-moving particle can lose energy by some interaction with an atom or molecule. An event in which such interaction occurs is frequently referred to as a collision, although it is difficult to give a rigorous definition of a collision which is appropriate in all cases. In the present context a collision may be defined as an encounter between particles in which there is a change in the internal energy or the kinetic energy or the momentum of the individual particles.

If a collision between two particles involves only an exchange of translational kinetic energy, the collision is said to be *elastic*. If the collision involves the interchange of some other form of energy, the collision is said to be *inelastic*. An inelastic collision, for example, might involve the interchange of internal energy of excitation or ionisation. Collisions often involve a change in the direction of motion of a particle, and the terms *elastic* and *inelastic scattering* are frequently encountered.

The result of a collision between two particles cannot always be predicted from a knowledge of the speed and direction of motion before impact. If, for example, an electron strikes an atom and the energy of the electron is less than the minimum excitation energy of the atom, then the collision can only be an elastic one. If, on the other hand, the energy of the electron is greater than the energy required to excite the atom, it is not certain that the atom will be excited as a result of the collision. It can be said only that there is a certain *probability* that the atom will be excited. The probability that a particular event will result from a collision in a given system is not a constant but varies with the energy of the colliding particles. There are, therefore, energy ranges in which certain results are more likely to occur than others. This will be appreciated from the later consideration of the mechanisms by which ionising particles transfer energy to the matter through which they pass.

The above remarks apply, in general, to both charged and neutral particles; for charged particles there is another way in which energy can be lost. If a charged particle passes through an electric field, it can be decelerated and the energy lost by this deceleration appears as electromagnetic radiation. This process does not necessarily involve an encounter between two particles and so does not come within the definition of a collision given above. Of course, when a charged particle passes through matter, it encounters the electric fields of atomic nuclei and can thus lose energy by radiation, but this process is not usually regarded as a collision. Once again, the probability of this process depends upon the energy of the charged particle.

The interaction with matter of the three classes of ionising radiation can now be considered in more detail.

CHARGED PARTICLES

There are three main processes by which charged particles interact with matter when passing through an absorbing medium. The probabilities, and thus the relative importance, of these processes depend very largely upon the energy of the incident particles, but there is also some dependence on the nature of the material through which they pass.

EMISSION OF RADIATION

When high-speed charged particles pass close to an atomic *nucleus* they may be decelerated, and the energy lost by this deceleration appears as electromagnetic radiation. This radiative loss is in accordance with classical physics and modern quantum theory, and is the bremsstrahlung described in Chapter 2.

The energy loss by bremsstrahlung emission per unit distance travelled by the charged particle, $-(dE/dx)_{rad}$, is given by:

$$-\left(\frac{dE}{dx}\right)_{rad} = k\frac{z^2Z^2}{m^2} \qquad (3.1)$$

where z and m are, respectively, the charge and mass of the particle, Z is the charge of the atomic nucleus retarding the particle and k is a constant. It can thus be understood from equation (3.1) that radiative loss is greatest for light particles travelling through a medium of high atomic number.

For electrons, bremsstrahlung emission is only appreciable at electron energies greater than 100 keV. The fraction of the electron's kinetic energy which is lost as bremsstrahlung emission increases rapidly with the actual energy of the electron. For 100 MeV electrons passing through water, about half of the energy is lost by radiation; and theory shows that at energies in excess of 150 MeV, radiative loss is responsible for most of the electron's energy loss.

Positive ions such as protons, deuterons, α-particles, etc., may also lose energy by bremsstrahlung emission, but in these cases the loss is only important for ion energies of about 1000 MeV and greater.

If the bremsstrahlung escapes from the absorbing medium, no significant changes are produced in it. If, however, the bremsstrahlung is subsequently absorbed by the medium, this is tantamount to irradiating the material with X-rays, the effects of which will be treated later.

INELASTIC COLLISIONS

If the energy of a high-speed charged particle moving through an absorbing medium is such that radiative loss by bremsstrahlung emission is negligible, then most of the kinetic energy is lost by electrostatic interaction with the *electrons* of the medium. When the moving particle approaches a molecule lying near to its path, the electric field of the charged particle affects the electrons of the molecule. Some of the kinetic energy of the particle is transferred through the electric field to the molecular electrons as internal energy. Owing to this gain of energy, some of the molecular electrons are promoted from the ground state to a higher energy level. This results in the formation of an excited state or, if the energy transfer is sufficient, the formation of an ion, the electron leaving the molecule altogether.

It will be appreciated that if the energy is transferred from the charged particle to a molecular electron through the electric field of the charged particle, then the stronger the electric field, the more chance there will be of energy transfer. The probability of energy deposition in the medium thus increases with the charge carried by the particle. Furthermore, the more slowly moving the particle, the greater the period of time during which the molecular electron is under the influence of the particle's field. The probability of energy transfer is thus greater for a more slowly moving particle.

Consider, for example, an electron and an α-particle having the same kinetic energy and travelling through the same medium. As the α-particle is much more massive than the electron, it will be travelling much more slowly. Moreover, the α-particle is doubly charged, whereas the electron is only singly charged. Under these circumstances it can be readily understood that the density of ionisation created by an α-particle will be much greater than that produced by an electron of the same energy. The density of ionisation and excitation which is produced will, of course, depend on the electron density of the medium. The greater this is, the greater will be the chance of an inelastic collision in a given volume and, hence, the greater will be the density of ionisation. The above illustration is, therefore, strictly valid only for a given medium.

As a charged particle travels through an absorbing medium, it transfers energy to the medium at the expense of its own kinetic energy. As it progresses, then, its speed is reduced and consequently the intensity of ionisation and excitation which it produces increases towards the end of its path. A maximum is usually shown just before the end of the path, after which the intensity of ionisation drops to zero when the particle no longer has sufficient energy to cause ionisation. This is illustrated in *Figure 3.1*, where the specific ionisation (see below) produced by α-particles is plotted against the distance travelled in air.

ELASTIC COLLISIONS

Elastic scattering occurs when a charged particle is deflected from its course by coulomb interaction with the electric field of an atomic nucleus. It is more important for electrons, as they have such a small mass that they are frequently deflected by such collisions. The atomic nucleus, on the other hand, is so massive that it can be considered to remain at rest during the scattering process. Elastic collisions are essentially a change in the direction of motion of the particle without any loss of its kinetic energy. Obviously, as it is the electric field of the atomic nucleus which is the effective agent, the greater the field, the

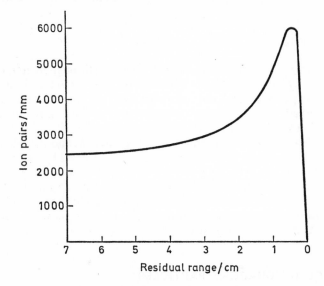

Figure 3.1.   Variation of ionisation with distance

more important is elastic scattering. Furthermore, slower particles will be deflected to a greater extent, so that elastic scattering is most important for low-energy particles travelling through a medium of high atomic number. In this case the atomic nuclei will be fairly massive even compared with α-particles, protons and deuterons, and may still be considered to remain at rest even in encounters with these heavy charged particles. While elastic scattering is quite important in the case of electrons on account of their small mass, for heavy charged particles it is relatively unimportant.

ENERGY LOSS BY CHARGED PARTICLES

Of the three processes described above by which a charged particle can interact with matter, only two reduce the energy of the particle.

These are the emission of radiation and inelastic collisions. The third mode of interaction—elastic collisions—does not change the kinetic energy of the particle but merely changes its direction of motion. The energy loss per unit path length due to radiation, $-(dE/dx)_{rad}$, and that due to inelastic collisions, $-(dE/dx)_{coll}$, are combined to give the total energy loss per unit path length, which is known as the *specific energy loss* or the *stopping power* of the absorbing medium.

DEPOSITION OF ENERGY IN MATTER

Of the energy lost by a charged particle passing through matter, only that lost by inelastic collisions is transferred directly to the absorbing medium, provided that any bremsstrahlung escapes without being re-absorbed. These inelastic collisions give rise to ionisation and excitation of the medium. An electron and the positive ion which results from an ionisation process form an *ion pair*, and the intensity of ionisation caused by a charged particle is expressed by the specific ionisation, which is the total number of ion pairs produced per millimeter of path length. The specific ionisation is thus a measure of the energy deposited in the medium, and a plot of the number of ion pairs formed against distance travelled, of the type shown in *Figure 3.1*, is known as a Bragg curve.

RANGE OF CHARGED PARTICLES

As pointed out above, charged particles lose their energy principally by inelastic collisions with the electrons of the medium through which they travel. These energy losses retard the particle until it no longer has sufficient energy to cause ionisation and excitation of the medium. The particle does not necessarily come to rest at this point, but it is no longer an effective agent in transferring energy to the medium and it is regarded as having come to the end of its path for all practical purposes. The path length of a charged particle in a medium may thus be considered as the distance it travels from the point of entry to the point where it no longer causes ionisation. There is, however, a difference in the shapes of the paths of electrons and heavy positive ions, which is due to the difference in their masses.

In an elastic collision between, say, an α-particle and an electron in the medium, the α-particle is so massive, compared with the electron, that it loses only a very small fraction of its energy and suffers little or no deflection. α-particles thus lose their energy in a large number of events involving the transfer of small amounts of energy and travel in very nearly straight lines. This can readily be seen from *Figure 3.2*, which is a representation of the tracks of α-particles observed in a cloud chamber.

Although a monoenergetic group of α-particles from a radioactive source have the same initial energy, they do not all lose the same amount of energy in their encounters with the molecules in their path. There is thus some 'straggling', and there are very slight differences

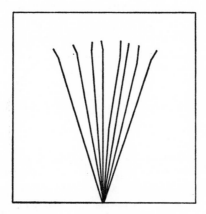

Figure 3.2. Alpha-particle tracks

in the path lengths of α-particles with the same initial energy. It will also be noticed that a very few α-particles are deflected from a straight line, especially towards the end of their paths. These deflections are due largely to elastic scattering, the probability of which increases as the energy of the particle decreases towards the end of its path. As stated previously, elastic scattering is relatively unimportant for α-particles. This can be appreciated from *Figure 3.2*.

Since α-particles travel in virtually straight lines, the depth of penetration into an absorbing medium is the same as the path length and is known as the *range* of the α-particle. Because all the α-particles from a single source do not travel exactly the same distance, however, reference is usually made to a *mean range* or an *extrapolated range*. The situation is illustrated in *Figure 3.3*, in which the number of α-particles from a given source is plotted against distance. This plot is a graphic representation of *Figure 3.2*. The section *ab* of the graph shows that the number of α-particles remains fairly constant with distance until the end of the path is approached. Section *bc* represents the decrease in numbers over the distance where straggling occurs. The extrapolated range, R, is found by extrapolating the falling part of the curve to the axis whilst the mean range, R_0, is the distance corresponding to the point of inflection of the section bc.

When the charged particles are electrons, the situation is somewhat different. In this case the inelastic collision involves an encounter between an electron (charged particle) and an atomic or molecular electron of the absorbing medium. The collisions are thus between two particles of the same mass; the incident electron can lose up to half its energy in such an encounter and may also be deflected

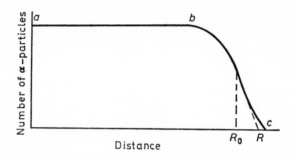

Figure 3.3.   Mean and extrapolated ranges

through a large angle. It will be remembered that elastic scattering is also important in the case of electrons, and large deflections will also occur in this way. As a result, monoenergetic electrons have no fixed depth of penetration into an absorbing medium but show a maximum distance of penetration, which is known as the *maximum range*. Unlike the case of α-particles, the maximum range of electrons is much shorter than the actual path length. For monoenergetic electrons the situation is represented in *Figure 3.4*, in which the number of electrons is plotted against the depth of the absorber which has been traversed. It can be seen that the number of electrons decreases continuously and very nearly linearly as the distance into the absorber increases, until it drops to a background level, which is due to bremsstrahlung.

With β-particles the position is somewhat different. It will be remembered that β-particles from a radioactive source are not monoenergetic but have a wide spectrum of energies. This fact, combined with the usual scattering effects just described, leads to an approximately exponential law for the absorption of β-particles of a given maximum energy. It is of the same form as the law which applies to

the absorption of electromagnetic radiation which is given in equation
(1.1), the number of β-particles at any point being proportional to the
intensity of the beam at that point. A plot of the logarithm of the
number of β-particles against the depth of absorber traversed is thus
approximately linear, an example is given in *Figure 3.5*. For both

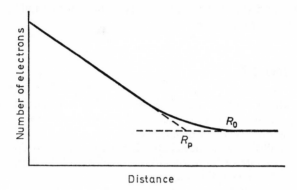

Figure 3.4.   Ranges of monoenergetic electrons

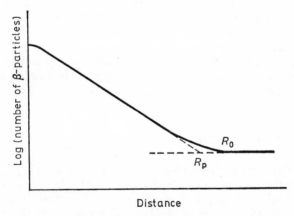

Figure 3.5.   Ranges of β-particles

monoenergetic electrons and β-particles the maximum range, R_0, is
taken as the point where the curve merges with the background, and
the extrapolated or practical range, R_p, is obtained by extrapolating
the approximately linear part of the falling curve to the background
level.

In connection with absorption of radiations by various materials, use is made of the product of the thickness of an absorber and its density to give a quantity in terms of mass per unit area. The ranges of charged particles are frequently expressed by this quantity where the range is multiplied by the density of the absorber. Ranges expressed in kilogrammes per square metre are relatively small for solid materials, and they are usually expressed in terms of milligrammes per square centimetre.

NEUTRONS

Since neutrons carry no electric charge, they do not interact with the electrons or atomic nuclei of an absorber through an electric field effect. The only way in which a neutron can interact with matter is by direct collision with the subatomic particles of the medium. Such collisions are infrequent, owing to the small size of the particles concerned, and neutrons can thus penetrate much greater thicknesses of material than charged particles before their energy is dissipated. Neutrons interact with matter almost exclusively by collisions with atomic nuclei and do not produce any ionisation directly. Neutrons are still included in ionising radiations, as the products of neutron interaction with nuclei can often cause ionisation and excitation.

There are four main processes by which neutrons can interact in collisions with atomic nuclei, and the type of interaction depends largely upon the energy of the neutron.

ELASTIC SCATTERING

For high-energy or *fast neutrons* an important process in which they transfer their energy to the absorbing material is by elastic collisions with atomic nuclei. This is simply an elastic impact between two bodies, and the energy of the impinging neutron is shared, after collision, between the neutron, which is deflected, and the nucleus. Elastic collisions are governed by the laws of conservation of momentum and kinetic energy. For elastic spheres the nature of the collision depends on the angle, θ, between the line of centres at impact and the direction of relative motion of the spheres. For example, consider a sphere of mass m_1 striking an initially stationary sphere of mass m_2, as illustrated in *Figure 3.6*. By application of the laws of conservation of momentum and kinetic energy it can be shown that the fraction,

δ, of the kinetic energy of m_1 which is transferred to m_2 as a result of the collision is given by

$$\delta = \frac{4m_1 m_2}{(m_1 + m_2)^2} \cos^2 \theta \qquad (3.2)$$

Thus the fraction of energy transferred varies from zero for a collision which is just a miss and $\theta = 90°$ to a maximum for a head-on collision, where $\theta = 0°$.

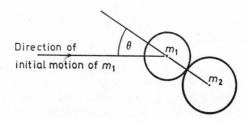

Direction of initial motion of m_1

Figure 3.6. Elastic collision between spheres

If this situation is applied to the elastic collision of a neutron with an atomic nucleus, the maximum value, $\delta(\theta = 0)$, of the fraction of the neutron's energy transferred to the nucleus depends on the mass of the nucleus which is struck, and is given by

$$\delta(\theta = 0) = \frac{4A}{(1 + A)^2} \qquad (3.3)$$

where A is the relative atomic mass of the nucleus and replaces m_2 in equation (3.2) and m_1 has been replaced by unity, the mass of a neutron on the same scale.

Examination of equation (3.3) shows that $\delta(\theta = 0)$ is a maximum when $A = 1$. There is thus a maximum transfer of energy from the neutron when it collides with a hydrogen nucleus. In organic media which contain a large proportion of hydrogen the vast majority of energy loss from neutrons by elastic scattering occurs through collision with hydrogen nuclei. In these collisions the protons are ejected from the molecule at high energies and give rise to ionisation and excitation, as described in the previous section on charged particles. The molecule which the proton leaves remains as a negatively charged free radical which will probably dissociate. This aspect of free-radical behaviour will be dealt with later.

Heavier atomic nuclei than protons can, of course, be ejected by

D

elastic collisions with neutrons. In such cases, however, because the nucleus will be much heavier than the neutron, the energy with which it is ejected, the recoil energy, will usually be quite small and the effects produced by the recoil nucleus will be unimportant.

The probability of energy transfer by elastic scattering depends to a large extent on the energy of the neutrons. For neutrons with energies in excess of 1 MeV most of the energy will be dissipated in this way, but the process is also quite important for neutrons with energies as low as 1 keV.

INELASTIC SCATTERING

For fast neutrons with energies in excess of a few hundred keV, energy transfer can occur through inelastic collision with a nucleus in which the nucleus is raised to an excited state. The incident neutron is absorbed by the nucleus which then re-emits a neutron of lower energy, the remaining nucleus being in an excited state. The nucleus eventually returns to the ground state, emitting the energy in the form of γ-rays of one or more quantum energies (i.e. wavelengths). These γ-rays can then give rise to ionisation and excitation in the manner described in the next section.

Obviously, the energy of the incident neutron must be at least equal to the difference between the ground state and the lowest excited state of the nucleus, and the importance of this process increases as the neutron energy increases. Inelastic collisions may be as probable as elastic collisions at neutron energies greater than 10 MeV.

NUCLEAR REACTIONS

A fast neutron colliding with an atomic nucleus may be incorporated in the nucleus, which then emits a different particle such as a proton or an α-particle. Such nuclear reactions only occur when the incident neutron has an energy greater than the threshold energy of the reaction and are only of any importance at energies above a few MeV. There are, however, some nuclear reactions which can occur with slow neutrons of thermal energies, such as $^{14}N(n, p)^{14}C$. The energy of the particle which is emitted will be equal to the energy of the incident neutron plus any energy liberated by the reaction. This may be positive or negative according to circumstances. If the emitted charged particles have sufficient energy, they can cause ionisation and excitation in the medium, as described previously.

CAPTURE

When a neutron has lost most of its kinetic energy in excess of that due to thermal motion, it is often called a *thermal neutron* or a *slow neutron*. Such neutrons do not have sufficient energy to eject particles from atomic nuclei but can be captured by a nucleus to give an isotope. The isotope is often formed in an excited state and returns to the ground state, emitting the excess energy in the form of one or more quanta of γ-radiation. This γ-radiation can then cause ionisation and excitation in the medium, as in the case of inelastic scattering.

Sometimes the Szilard–Chalmers effect can be observed in neutron-capture reactions. If a molecule is represented by RA^x, where A^x is an atom of mass x and R is the rest of the molecule, a neutron may be captured by A^x to give an isotope, A^{x+1}. A γ-ray is emitted and, in order to conserve momentum, the isotope A^{x+1} acquires a recoil energy. This energy will be sufficient to break a chemical bond; and if it is not spread over the whole molecule, the isotope may be ejected from the parent molecule with a high kinetic energy:

$$RA^x + {}^1n \longrightarrow RA^{x+1} + \gamma$$
$$RA^{x+1} \longrightarrow R\cdot + A^{\cdot x+1}$$

The remainder of the molecule is left as a free radical and the kinetic energy of the isotope may be sufficient to break other chemical bonds in neighbouring molecules. Such reactions are illustrated by the example of methyl iodide:

$$CH_3{}^{127}I + {}^1n \longrightarrow CH_3{}^{128}I + \gamma$$
$$CH_3{}^{128}I \longrightarrow CH_3^{\cdot} + {}^{128}I$$

In this case the ^{128}I can react with other methyl iodide molecules.

Neutron capture is the more likely to occur, the longer the neutron remains in the vicinity of the nucleus; it is thus most likely at low neutron energies. As the neutron energy increases, the probability of capture diminishes but can show peaks at certain resonance energies. For fast neutrons the probability of capture is low.

In organic media or aqueous solutions nearly all slow neutrons are captured by hydrogen atoms forming deuterium with the emission of a 2·2 MeV γ ray, although, if nitrogen is present, the nuclear reaction mentioned in the previous section can occur with slow neutrons.

CROSS-SECTIONS

It was mentioned at the beginning of this chapter that the result of an interaction between a particle and matter can only be expressed in terms of a probability. These probabilities are frequently stated in terms of an *effective cross-section* for a particular event.

If a ball-bearing of radius r_1 is allowed to fall on a table which supports a circular glass plate of radius r_2, then the ball-bearing will strike the plate if its centre falls anywhere within a circle of radius $(r_1 + r_2)$ from the centre of the plate. The cross-sectional area within which the centre of the ball-bearing must fall in order to strike the plate is thus $\pi(r_1 + r_2)^2$, and this is known as the cross-section for the collision. It can be understood from this that the greater the cross-section, the greater is the probability of a collision.

Now suppose that there are several plates of the same size resting on the table and that a continuous stream of ball-bearings is allowed to fall on the table. If there are n plates, each of collision cross-section σ, then the total surface in which collisions can occur is $n\sigma$. If the area of the table is A, then the fraction of the table area in which collisions occur is $n\sigma/A$. Thus, if the rate at which ball-bearings fall on the table is R_0, then the rate at which collisions occur, R, is given by

$$R = R_0 \frac{n\sigma}{A} \qquad (3.4)$$

The cross-section for collision could be calculated from equation (3.4), as R, R_0, n and A can all be measured. Hence,

$$\sigma = \frac{RA}{R_0 n} \qquad (3.5)$$

The concept of cross-section is not restricted to collisions. For example, if some of the plates were broken by the impact at a rate R_b, then the cross-section for breakage, σ_b, could be calculated from

$$\sigma_b = \frac{R_b A}{R_0 n} \qquad (3.6)$$

If the probability of breakage is high, then R_b will be large and thus σ_b will be large. It will be appreciated that the cross-sections for particular events bear no relation to the geometrical area of the bodies involved but simply express the probability of a particular event.

The probabilities of the results of interaction of particles with matter can be expressed in terms of cross-sections and on the atomic scale they often have values of the order of 10^{-28} m^2 atom^{-1}. For this reason a cross-section of 10^{-28} m^2 has been called 1 *barn*, but this term is no longer allowed in the SI system.

ELECTROMAGNETIC RADIATION

When one considers the interaction of electromagnetic radiation with matter, it is probably best to regard the radiation as a stream of photons of quantum energy E, where

$$E = h\nu \tag{3.7}$$

h being Planck's constant and ν the frequency of the radiation. The intensity of the radiation may now be regarded as the number of photons passing through unit area in unit time. The behaviour of these photons is rather different from the interactions of charged particles and neutrons, where the behaviour can be characterised by a depth of penetration or a range. In these cases a particle loses energy in a large number of successive events and is gradually slowed down until there is no further interaction. When a beam of photons falls on a thin absorber, however, photons are removed from the beam completely in single events. A photon may be absorbed, depositing its quantum of energy in the material, or it may be deflected out of the beam. In the latter case it may be deflected with no significant change of energy or it may be deflected with a loss of energy, in which case it will appear as radiation of a longer wavelength.

Photons have no definite range; they may be removed from the beam after travelling only a short distance or they may penetrate a long way. The greater the distance travelled by a photon in an absorber, the greater is the probability of its being absorbed. As photons are removed from the original beam, the intensity must decrease and the beam is said to have been attenuated. This attenuation is the result of the combined effects of outright absorption and deflection, although the words attenuation and absorption are sometimes used synonymously.

Because photons are removed from the beam in single events, the intensity of electromagnetic radiation is related to the distance travelled in an absorber by an exponential law. Consider a beam of photons incident on an absorber of thickness x as illustrated in *Figure 3.7*. Suppose that the intensity of the incident beam is I_0 and that, when the beam reaches the thin section of absorber of thickness dx, the intensity has been reduced to I. The decrease in intensity $-dI$ on passing through the section of thickness dx will be proportional to dx. Furthermore, the more photons there are in the beam, the more likely is interaction to occur. Thus $-dI$ will also be proportional to I, or

$$-dI = \mu I dx \tag{3.8}$$

where μ is a proportionality constant. Integrating equation (3.8) for the total thickness of the absorber,

$$I = I_0 e^{-\mu x} \qquad (3.9)$$

For a given absorber and a given wavelength of radiation μ is a constant and is known as the *linear absorption coefficient*. As the exponent in equation (3.9) must be dimensionless and as x has the dimensions of length, the dimensions of μ must be reciprocal length.

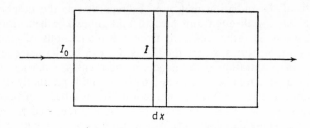

Figure 3.7. Absorption of electromagnetic radiation

From equation (3.9) it might be expected that the intensity of the radiation would never become zero no matter how thick the absorber. It is, however, doubtful whether the equation is valid at very low intensities, because the considerations upon which equation (3.8) is based will apply only to a large number of photons, as it is concerned with probabilities. In practice the radiation is undetectable beyond a certain point.

It is found experimentally that the absorption of radiation depends only on the number of atoms in the absorber and is independent of their state of aggregation. Thus, if the linear absorption coefficient, μ, is divided by the density, ρ, of the material, a mass absorption coefficient, μ/ρ, is obtained which is independent of any changes in density due to the physical state of the absorber. In this case equation (3.9) is best expressed as

$$I = I_0 e^{-\mu m/\rho} \qquad (3.10)$$

where m is the mass per unit area of the absorber. (Compare this method of expressing the distance travelled in an absorber with the method of expressing the range of a charged particle.) The relation is expressed in this way in order that the exponent shall be dimensionless. Other useful expressions of absorption coefficients are as atomic absorption coefficients, $_a\mu$, or electronic absorption coefficients, $_e\mu$, which are equal to the linear absorption coefficient divided by the number of atoms or electrons per unit volume of absorber, respectively. The relationships are thus given by

$$_a\mu = \frac{\mu A}{\rho L} \tag{3.11}$$

$$_e\mu = \frac{\mu A}{\rho L Z} \tag{3.12}$$

where A is the molar mass of the atoms, L is Avogadro's constant, ρ is the density and Z is the atomic number.

The mass, atomic and electronic absorption coefficients involve the dimensions of area and are frequently called cross-sections, as they refer to the probability of absorption. As atomic and electronic absorption coefficients are often of the order of 10^{-28} m^2 per atom or electron, they have frequently been expressed in terms of barns per atom or electron.

There are three important processes by which electromagnetic radiation interacts with matter. The relative importances of these three vary with the energy of the radiation, and the probability of each may be expressed in terms of an absorption coefficient relating to the particular process. These processes are the photoelectric effect, the Compton effect and pair production, the individual absorption coefficients being denoted by the symbols τ, σ and κ, respectively. The total linear absorption coefficient μ, is thus the sum of the three individual absorption coefficients:

$$\mu = \tau + \sigma + \kappa \tag{3.13}$$

THE PHOTOELECTRIC EFFECT

Low-energy photons may be completely absorbed by transferring the whole of their energy to a single atomic electron. As a result the electron is ejected from the atom with a kinetic energy equal to the difference between the energy of the incident photon and the binding energy of the electron in the atom. The ejected electrons can then cause ionization and excitation in the medium, as described in the section on charged particles. The energies of the electromagnetic radiation encountered in radiation chemistry are usually sufficiently high to remove one of the most tightly bound electrons in the atom from the K-shell. As a result of this vacancy an electron from an outer shell can fall into the K-shell, and this process is accompanied by characteristic X-ray emission or, more rarely, by the ejection of one of the other electrons in the atom. The energy of these secondary

X-rays or electrons (known as Auger electrons) will be equal to the energy lost by the atomic electron which falls from an outer shell to fill the vacancy in the K-shell left by the photoelectron. In absorbers of low atomic number the binding energy of the K-shell is small, and thus the secondary X-rays and electrons will only have low energies and will be absorbed locally.

In the photoelectric effect the direction taken by the ejected primary electron depends to some extent upon the energy of the absorbed photon. Most electrons are ejected perpendicularly to the direction of motion of the photon; but as the photon energy increases, the photo-electrons become increasingly scattered in the forward direction. As momentum must be conserved in these events, the atom must recoil in a different direction to that in which the electron is ejected. It can thus be appreciated that this type of photoelectric effect cannot be observed with free electrons.

THE COMPTON EFFECT

If the binding energy of an electron in an atom is low compared with the energy of the incident photon, it may be regarded as a free electron. If the photon collides with a free or loosely bound electron, it

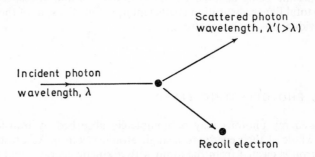

Figure 3.8. The Compton effect

may impart some of its energy to the electron, which will move off with a kinetic energy depending on the nature of the collision. The photon continues in a different direction with a lower energy. The Compton effect thus provides a recoil electron and scatters the incident radiation at a longer wavelength. The situation is represented in *Figure 3.8*. The recoil energy of the electron is equal to the difference in energy between the incident and the scattered photon. This can be a range of values from zero (when the photon just misses the electron) to a maximum when the direction of the photon is completely reversed.

The Compton effect can be regarded from different points of view. If the incident electromagnetic radiation is monochromatic, then all the incident photons have the same energy. As a result of Compton interaction, photons of this energy disappear. There is, then, an absorption coefficient corresponding to the disappearance of photons of the original energy. This is called the *total Compton absorption coefficient* and is denoted $_e\sigma$. Some of the original energy, however, is retained by the scattered photons of lower energy and there is another absorption coefficient, the *scattering coefficient*, $_e\sigma_s$, which corresponds to the fraction of the original photon energy which is retained by the scattered photons. If interest centres on the fraction of the original energy which is transferred to electrons, this is the difference between the energy of the incident and the scattered photons, and the corresponding absorption coefficient is the *energy Compton absorption coefficient*, $_e\sigma_a$, which is equal to the difference between the first two absorption coefficients:

$$_e\sigma_a = {}_e\sigma - {}_e\sigma_s \qquad (3.14)$$

Compton interaction is favoured over a range of energies which has a higher and a lower limit. For water, Compton interactions predominate in the energy range 30 keV–20 MeV. Although the ejection of an electron from an atom leaves an ion, it is the ejected electron which goes on to cause most of the ionisation and excitation due to this type of interaction of γ-rays and X-rays. The scattered photons of lower energy can continue to interact with the absorber provided that they have retained enough of the incident photon energy.

PAIR PRODUCTION

Einstein's theory of the equivalence of mass and energy may be expressed by the equation

$$E = mc^2 \qquad (3.15)$$

where E is energy, m is mass and c is the velocity of electromagnetic radiation. According to equation (3.15), the mass of an electron is equivalent to 0·51 MeV. The electron is, of course, negatively charged, but the production of a negative charge can only occur with the production of an equal positive charge. It thus appears that the production of an electron requires the production of an equal mass of positive charge. This has been called a *positron*, and the amount of energy needed for the production of an electron–positron pair will be 1·02 MeV. If a photon of energy greater than 1·02 MeV is absorbed in the vicinity of an atomic nucleus, the result is the production of an electron and a positron. The energy of the photon in excess of 1·02 MeV is shared largely by the electron and the positron, and momentum is conserved by the recoil of the nucleus. Both the electron and

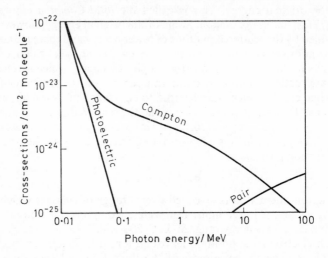

Figure 3.9.   Cross-sections for photon interactions

the positron can interact with matter in the usual manner of charged particles. The positron, however, will soon encounter another electron, and the two will suffer mutual annihilation with the creation of an equivalent amount of energy. As there are many electrons in matter, the positron exists only for a very short time (about 10^{-9} s). The energy created is called *annihilation radiation* and consists of two photons, each of quantum energy 0·51 MeV moving in opposite directions. These photons can then interact with matter by the photoelectric effect or the Compton effect, as described above.

The cross-sections of the three processes, the photoelectric effect, the Compton effect and pair production are shown as a function of photon energy in *Figure 3.9* for water. There are other processes by which photons can be removed from the incident beam but their cross-sections are so low in the energy range of ionising radiations that they need not be considered here.

EFFECTS OF IONISING RADIATIONS UPON MATTER

At the beginning of the chapter it was stated that radiation chemistry is the study of the reactions which result from the deposition in matter of the energy of ionising radiations. The various modes of interaction of radiation with matter having been dealt with, it will now be useful to summarise the processes by which the different types of radiation transfer energy to the medium through which they pass.

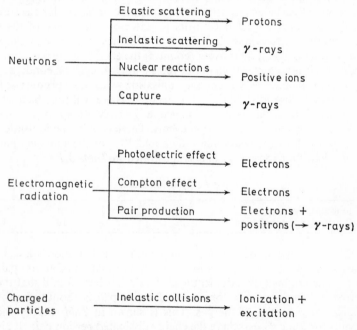

Figure 3.10.   Summary of interactions

An over-all idea of the processes involved can be obtained from *Figure 3.10*, in which the net results of the interactions described in this chapter are represented.

Neutrons produce γ-rays by inelastic scattering and capture and also produce protons and other positive ions by elastic scattering and nuclear reactions. The γ-rays resulting from neutron interaction behave in the ways represented for electromagnetic radiation, all three of which produce electrons. The positrons resulting from pair production quickly produce annihilation radiation in the form of γ-rays, which will then interact by the photoelectric effect or the Compton effect to produce electrons. The effects of neutrons and electromagnetic radiation are thus the eventual production of charged particles. It will be recalled that charged particles transfer energy to matter by inelastic collisions which result in ionisation and excitation.

The over-all effects of all three types of ionising radiation are thus the creation of ionisation and excitation by the action of charged particles in the medium through which they pass. Further consideration can now be given to these ionisation and excitation processes.

An ionisation event occurs when enough of the kinetic energy of the charged particle is transferred to an atomic or molecular electron in the medium to eject it from its parent molecule. Some of these secondary electrons may be ejected with sufficient energy to cause ionisation events on their own account. If the energy of the secondary electrons is not greater than about 100 eV, their range in a condensed medium will be very short and any ionisation which they produce will be situated close to the original ionisation. There will thus be a site of dense ionisation, which is called a *spur*. Most spurs contain between one and four ionisation events. There is only one ionisation in about half of the spurs and only about 10% of the spurs contain more than four ion pairs. This is illustrated in *Table 3.1*

Table 3.1 Spur ionisation

Ionisations per spur	1	2	3	4	>4
Percentage of spurs	44	22	12	10	12

The spurs will, of course, also contain excited molecules. In gaseous media the energy dissipated for the formation of an ion pair can be determined directly. In these cases it has been found that the energy consumed for the formation of an ion pair is about twice the ionisation potential of the gas. This is shown in *Table 3.2* by the results for some gases, where the energy dissipated per ion pair, W, is given together with the ionisation potential, I, for a few gases.

Table 3.2 Ion pair energies and ionisation potentials

Gas	W/eV	I/eV	W/I
Hydrogen	36·31	15·38	2·36
Helium	42·31	24·50	1·73
Nitrogen	34·68	15·62	2·22
Oxygen	30·87	12·31	2·51
Methane	27·31	13·12	2·08

The excess energy is used in causing excitation, so that approximately equal amounts of energy are consumed in the ionisation and excitation processes.

In condensed systems it is not possible to measure accurately the number of ions produced by a given amount of energy. Some attempts have been made with liquids, and the results suggest that the energy dissipated per ion pair in these cases is about the same as that found for the corresponding gases. It is generally assumed that W does not depend upon the physical state of the medium.

Secondary electrons with energies in excess of a few hundred eV will have a greater range than those secondary electrons which give

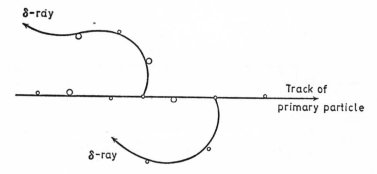

Figure 3.11.   Delta-ray tracks

rise to spurs. They will thus form tracks of their own as a branch to the track of the original charged particle. Electrons which do this are called δ-*rays*. It is believed that about 60–80% of the ionisations caused by a primary α-particle are due to the action of secondary electrons and for a primary electron about 70–80% of the ionisations are due to secondary electrons. It has also been calculated that half the ionisation events caused by any primary charged particles are to be found in δ-ray tracks.

The general situation which has been described above can be represented schematically as in *Figure 3.11*, in which the circles of different sizes represent spurs containing different numbers of ion pairs and excited molecules. It will be appreciated that if the primary particle is a positive ion, its track will be virtually linear; but if it is a fast electron, many deviations will occur owing to frequent scattering. Moreover, it will be recalled that positive ions give rise to a much greater density of ionisation than do electrons in the same absorbing medium. Thus, if the primary particle is a positive ion, the spurs will be formed more closely to one another than if the primary particle were an electron. In condensed media the spurs from a positive ion track are formed so closely to one another that they overlap, forming a cylindrical column of ions and excited species along the primary track. On the other hand, if the primary particle is an electron, the spurs will be spread out at much greater intervals. In *Figure 3.12* the electron track is drawn as linear for simplicity.

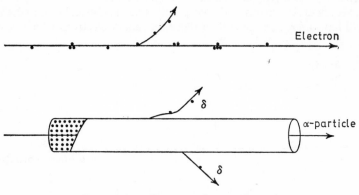

Figure 3.12. *Electron and α-particle tracks*

As all ionising radiations give rise to ions and excited species through the action of charged particles, the chemical effects of the various ionising radiations are qualitatively similar. The differences in the linear density of events along the track of the charged particle, however, cause quantitative differences in behaviour, which will be discussed in detail in later chapters. It is thus useful to have some measure of the linear density of events and this is given by the *linear energy transfer* (LET), which is defined as 'the linear rate of loss of energy (locally absorbed) by an ionising particle traversing a medium'.

It will be appreciated that the LET of a particular type of radiation is not constant along the whole of its path. For example, *Figure 3.1* shows that the ionisation produced by an α-particle is greatest towards the end of its path and, hence, the LET of the particle must also be greatest in this region. It must also be remembered that the density of ionisation produced by a given charged particle depends on the electron density of the medium. Thus, for any given type of particle, the LET will be much less in a gas than in a condensed medium.

Usually an average value of LET of a type of radiation in a given medium is quoted and a rough value can be obtained by dividing the total energy of the particle by its path length. Better average LET values may be obtained by more sophisticated calculations, which can account for the fact that the energy transferred to a δ-ray is not necessarily absorbed locally and which may also allow for the wide distribution of LET of the secondary electrons, which are produced with a wide range of energies by electromagnetic radiations. Because of the different methods by which such calculations are performed, care must be exercised in comparing LET values from different sources.

No matter how the LET values are calculated, for a given medium the minimum values are obtained for γ-rays, fast electrons and high-energy X-rays. They increase through the values for β-particles, protons and neutrons, deuterons, α-particles to the maximum values, which relate to heavy ions and fission fragments from nuclear reactions.

The differences in the quantitative results of various types of radiations are less marked in gases than in condensed media. The density of ions and excited species formed initially along the track of a charged particle and in the spurs very soon decreases as a result of diffusion into the bulk of the medium and by chemical reactions. Diffusion from the tracks is much easier in gases than in liquids, where the active species are 'caged' to a greater extent by the closely packed molecules in the liquid. The quantitative results of, say, α-particles and γ-rays are thus not very different for gaseous systems, in spite of differences in LET. In solution, however, the chemical action of γ-rays on the solute can be three times as great as that of α-particles for the same amount of energy absorbed.

4

Behaviour of ions, excited states and free radicals

The effect of any type of ionising radiation on matter is the production of ions and excited states. These processes may be represented:

$$A \longrightarrow A^+ + e^- \qquad (4.1)$$

$$A \longrightarrow A^* \qquad (4.2)$$

where A is a molecule of the absorbing medium, A^* is an excited state of the molecule and the symbol $\longrightarrow$ is used to denote a process brought about by ionising radiation. It has been shown in the previous chapter that the electrons produced in reaction (4.1) can cause further ionisation and excitation on their own account in spurs or δ-ray tracks if they are ejected from the parent molecule with sufficient energy. As such electrons cause further ionisation and excitation, their own energy decreases until a stage is reached where their energy is insufficient to cause ionisation and excitation and they have only thermal energies and are thus called 'thermal' or 'slow' electrons. The primary effect of ionising radiation is thus the production of ions, excited states and slow electrons. These species subsequently react in a number of ways, the result of some of which is the production of free radicals. The great majority of chemical changes induced by radiation, especially those occurring in the liquid phase, have been interpreted fairly successfully on the basis of free radical reactions together with the reactions undergone by thermal electrons.

There are indications, however, that for the explanation of some finer points of detail in radiation chemistry the reactions of excited states may have some importance.

It is thus necessary to have some knowledge of the behaviour of ions, excited states and electrons, first from the point of view of free radical production and second because the reactions of these species may have some importance in their own right to the study of radiation chemistry.

DETECTION AND STUDY OF IONS

The formation of ions in gases by ionising radiations can be readily observed in cloud chambers and can also be detected in ionisation chambers, the principle of which has been described in Chapter 1. One of the most powerful tools in the study of ions, however, is the mass spectrometer, and the majority of the currently available information on ions and their behaviour comes from gas phase studies in mass spectrometry. In the mass spectrometer, the gas to be examined is ionised by allowing slow electrons from an electron gun to impact with the gas molecules. The ions so formed are then accelerated by an electric field, after which they enter a magnetic field, which causes them to move in a circular path. The radius of this circular path depends upon the accelerating potential, the magnetic field strength and the ratio of the mass to the charge of ion. A plate containing a narrow slit is placed in the path of the ions, and for a fixed value of the accelerating potential and magnetic field strength only ions of a given mass to charge ratio will fall on the slit in the plate. Ions of other mass-to-charge ratios will have paths of different curvatures and thus will not fall on the slit. Those ions which pass through the slit are then collected by a detector and can subsequently be identified. In further observations, other ions can be made to arrive at the slit by changing the accelerating potential. In this way all the ions can be detected.

Usually the pressure in a mass spectrometer is very low, somewhere between 10^{-5} and 10^{-7} Torr (c. 1–0·01 mN m^{-2}) so that the ions, once formed, have little chance of undergoing a collision which might result in a reaction. Under these conditions, most of the ions which are collected can only have been formed by direct electron impact. Information on the reactions of ions with uncharged molecules can be obtained by coupling two mass spectrometers. The ions produced in one spectrometer are allowed to undergo ion–molecule reactions and the ionic products of these reactions are analysed by the second spectrometer.

E

Another method of interest to radiation chemistry is so-called 'high-pressure mass spectrometry'. A preliminary ionisation chamber operating at pressures up to 1 Torr (133 N m^{-2}) produces ions which can react with other molecules and the neutral species so formed are fed into the ionisation chamber of the mass spectrometer for analysis.

The low pressures which have to be used in mass spectrometers are rather a disadvantage from the point of view of radiation chemistry. Care must be exercised in applying the results of mass spectrometry to the radiation chemistry of gases at higher pressures, and it will be realised that the situation obtaining in the radiation chemistry of condensed media will be even further removed from that in a gas at low pressure. In the mass spectrometer an ion is virtually isolated once it is formed and any reaction which it subsequently undergoes can only be unimolecular. In gases at higher pressures, and especially in liquids, an ion will have a much greater probability of collision with the surrounding medium. This in turn gives rise to the possibility of bimolecular reaction of the ion with an adjacent molecule, and the products of such reaction could be very different from the result of a unimolecular reaction in the gas phase. In spite of this disadvantage, mass spectrometry provides more or less the only source of data on ion behaviour, and the results of mass spectrometry can be used, with some reservation, as a guide to the likely events in radiation chemistry.

REACTIONS OF IONS

The reactions of ions fall into three classes: neutralisation, ion–molecule reactions and charge transfer reactions.

NEUTRALISATION

Positive ions are eventually neutralised by either a slow electron or a negative ion. If an electron is the neutralising agency, the neutral species so formed will have gained an energy equal to its ionisation energy and will thus be in a highly excited state:

$$A^+ + e \longrightarrow A^{**} \tag{4.3}$$

If this process occurs in a gas at low pressure, the neutral entity may spontaneously re-ionise. At higher pressures, or in condensed phase where collisions can occur, some of the energy of the highly excited state can be dissipated and the neutral entity can achieve a lower excited state.

If the positive ion is also a radical the neutralisation by an electron provides a highly excited radical which is much more reactive than a normal radical and is sometimes called a *hot radical*:

$$R^{\cdot +} + e^- \longrightarrow R^{\cdot **} \qquad (4.4)$$

Neutralisation of a positive ion by a negative ion usually gives two neutral entities where either one or both may be in an excited state:

$$A^+ + B^- \longrightarrow \begin{cases} A^* + B \\ A^* + B^* \end{cases} \qquad (4.5)$$

ION–MOLECULE REACTIONS

In the early days of radiation chemistry, when the study was mostly confined to gases, ions were considered to be responsible for most of the chemical change. Later, when the studies were extended to liquids, ion–molecule reactions were largely ignored, as it was believed that in liquids neutralisation was a very fast process and occurred before ion–molecule reactions could take place. The reactions of ions were considered to be important only in gases where neutralisation was slow owing to the low concentration of ions or in solids where neutralisation was restricted by the low mobility of the ions. More recently, however, it has been realised that ion–molecule reactions can occur very rapidly and more attention is being turned to them. Little information is currently available on these reactions, but they seem to be mainly of two types. One type of reaction involves the transfer of a hydrogen atom from a molecule to an ion of the same type of molecule to produce a different ion and a free radical:

$$RH^+ + RH \longrightarrow RH_2^+ + R^{\cdot} \qquad (4.6)$$

The other type of reaction is a condensation reaction which may be represented:

$$A^+ + CD$$
or
$$\longrightarrow AC^+ + D$$
$$A + CD^+$$

in which D may be either a stable molecule or a free radical.

CHARGE TRANSFER REACTIONS

In these reactions an ion transfers its charge to a neutral molecule:

$$A^+ + B \longrightarrow A + B^+ \qquad (4.7)$$

For this reaction to occur at ordinary temperatures the ionisation energy of B must be not greater than that of A. This process is favoured when the difference between the ionisation energies of A and B is small. Any excess energy liberated in the reaction is usually retained by the products. If the energy released is sufficient it may cause the dissociation of the ion B^+ into a radical ion and a radical:

$$B^+ \longrightarrow C^{\cdot +} + D^{\cdot} \tag{4.8}$$

REACTIONS OF ELECTRONS

SIMPLE CAPTURE

In addition to neutralising positive ions as in reaction (4.3), electrons can become attached to neutral molecules to form negative ions:

$$A + e^- \longrightarrow (A^-)^* \tag{4.9}$$

This process is most likely to occur when molecule A has a large electron affinity such as oxygen, the halogens and halogen-containing organic compounds. In such a process an amount of energy corresponding to the electron affinity of A and the energy of the electron is released and this must be accommodated by the negative ion which is formed. This type of capture can occur only when the sum of the electron affinity and the electron energy corresponds to a difference in the energy levels of the negative ion. Such a process, where the probability of an event is greatest for certain energy values, is called a *resonance process*. It can be understood then that the negative ion will be formed in an excited state which can be stabilised by the emission of radiation (radiative attachment),

$$(A^-)^* \longrightarrow A^- + h\nu \tag{4.10}$$

or alternatively a third body (three-body attachment) can carry away the excitation energy of the negative ion.

DISSOCIATIVE CAPTURE

If the electron affinity of a molecule, due to the presence of an electronegative atom or group, is greater than the bond strength between that group and the rest of the molecule, the capture of an electron may immediately result in dissociation. This process may be represented

$$A + e^- \longrightarrow B^- + C^{\cdot} \tag{4.11}$$

In such a process C is usually a free radical and B^- may be a free radical ion. Once again, this process is most likely when B has a high electron affinity and will thus be most probable if B is a halogen or cyano group.

It should be remembered that the electron capture processes described above will compete with the neutralisation process (4.3). In condensed media, where the positive ions and electrons formed in the primary ionisation events along the track of a charged particle are hampered from diffusing, the neutralisation process will predominate in the early life of the track.

SUB-EXCITATION ELECTRONS

Electrons which have energies less than the lowest excitation energy of the medium are known as *sub-excitation electrons*. These electrons are believed to lose their energy more slowly than faster electrons, and thus have rather longer lifetimes. They will normally be neutralised by a positive ion or captured by a neutral molecule. They may become important, however, if the medium contains a low concentration of a substance with a lower excitation energy than the medium itself. Under these circumstances, the sub-excitation electrons cannot excite the molecules of the medium but may live long enough to encounter a molecule of the solute which could become excited. In this way there could be a larger number of direct excitation events involving the solute than might be expected on the basis of its concentration.

SOLVATION OF ELECTRONS

If, in a liquid medium, an electron escapes recombination with a positive ion at an early stage in the life of the track and also if its reaction with the medium is slow, it may become solvated. This will stabilise the electron and it may survive long enough to react with a solute. This is an important process in aqueous solutions and for media such as alcohols, where ion solvation is possible.

DETECTION AND STUDY OF EXCITED STATES

The production of excited states by ionising radiation has been inferred largely from indirect evidence. It has already been stated that the energy dissipated in the formation of an ion pair is in excess of the ionisation energy in most gases and the excess energy consumed is considered to be used in the formation of excited states. In addition, many of the reactions in radiation chemistry lead to the same products as those arising from photochemistry, where the reaction is initiated by the formation of an excited state. Much of the information on the existence and behaviour of excited states comes from spectroscopic and photochemical studies coupled with the application of theoretical chemistry. Ionising radiations produce the same excited states as those in photochemistry, but, in addition, they may produce more highly excited states than those formed by the absorption of light.

REACTIONS OF EXCITED STATES

The representation of the behaviour of the excited states of polyatomic molecules can be quite complicated, but an understanding of the processes involved can be gained by considering the properties of the excited states of simple diatomic molecules. Only the energy due to electronic energy and vibrational energy will be considered; any small contribution from rotational energy will be ignored. The potential energy of a diatomic molecule may be represented for any given electronic state by a curve giving the variation of potential energy with the separation of the two nuclei. *Figure 4.1* shows such curves for two electronic energy levels of a molecule, the lower curve, G, representing the electronic ground state and the upper curve, E, corresponding to the first electronic excited state. The horizontal lines within the wells of the curves represent the different vibrational states within each electronic state. The vertical distance from the lowest vibrational level of a particular state to the plateau of the curve where the nuclei are separated by an infinite distance is the dissociation energy of that state. The normal state of a molecule in a given electronic state is the lowest vibrational level of that state. A molecule which has a higher vibrational energy can easily lose this excess by transferring it as translational energy to another molecule in a collision.

When a molecule undergoes promotion from the ground state to an excited state, the process occupies about 10^{-15} s. The normal period of vibration of a molecule is about 10^{-13} s, so that during the excitation process the separation between the two atomic nuclei hardly changes and the transition can be represented by a vertical line on the potential energy diagram. A vertical line thus corresponds to the most probable transition and this is known as the *Franck–Condon principle*. Quantum-mechanical considerations predict that the most probable separation of the nuclei in the lowest vibrational level corresponds to the minimum in the potential energy curve, so that the most likely excitation process is represented on the diagram by the vertical line drawn from the middle point of the lowest vibrational level of the ground state.

The minima of the potential energy curves correspond to the

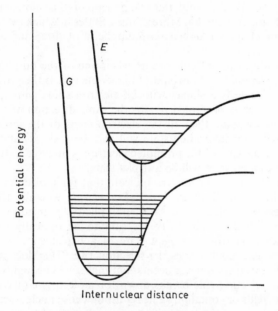

Figure 4.1.   Potential energy curves

equilibrium separation of the nuclei and very often this equilibrium separation is greater in an excited state than in the ground state. This is shown by the fact that the minimum of the excited state curve is to the right of the minimum of the ground state curve. As a result of this, promotion from the ground state with zero point vibrational energy can lead to an excited electronic state in which the molecule has a greater vibrational energy than the zero point energy for that state. Such a molecule will soon lose the excess vibrational energy by collisions and arrive at the lowest vibrational level of the excited state.

Most molecules have an electronic configuration in the ground state in which the spin of each electron is opposite to, or paired with, the spin of another electron. The total spin angular momentum of the electrons is thus zero, and the component of the spin angular momentum in a specified direction can only have the single value of zero. For this reason, states in which all the electron spins are paired are known as *singlet states* and are said to have a *multiplicity* of one. Any molecule with an even number of electrons, whether the spins are paired in the ground state or not, can have excited states where the spins of two electrons are parallel or unpaired as opposed to opposite or paired. If the spins of two electrons are unpaired, the net spin angular momentum will not be zero and the component of this momentum in a specified direction can have three values. States in which two electrons have parallel spins thus have a multiplicity of three and are called *triplet states*.

Excitation of a molecule occurs when one of the outermost electrons is promoted to an orbital of higher energy. If the ground state of the molecule is a singlet with all electron spins paired, and the electron which is promoted retains the same direction of spin in its higher energy state, then all the electron spins of the molecule are still paired and the excited state will also be a singlet state. Alternatively, if the spin of the promoted electron is inverted in the excitation, the excited state will be a triplet state.

Studies of electronic spectra indicate that for molecules not containing heavy atoms there is a *selection rule* whereby transitions occur between states of the same multiplicity. Transitions between states of different multiplicity are said to be optically forbidden. Thus transition from a singlet state to a triplet state is optically forbidden as it contravenes the selection rule. This does not mean that such a transition never occurs, but that it will be highly improbable. The selection rule breaks down for molecules either in strong magnetic fields or containing a heavy atom whose nucleus exerts such a field. In general, however, if the ground state of a molecule is a singlet, then excitation will normally be to another singlet state. Such a transition due to ionising radiation could be represented

$$^1A \longrightarrow\!\!\sim\!\!\sim\!\!\sim\!\!\rightarrow {}^1A^* \qquad\qquad (4.12)$$

Usually the lowest excited triplet state of a molecule has a lower energy than the lowest excited singlet state. Thus, although the most probable excitation process from a singlet ground state is to a singlet excited state, if the incident particle is a slow electron it may not have enough energy to excite the lowest singlet state of the molecule. It may, however, have sufficient energy to excite the lowest triplet state and it is believed that excited triplet states can be produced in this way in radiation chemistry:

$$^1A \xrightarrow[\text{electron}]{\text{slow}} {}^3A^* \qquad\qquad (4.13)$$

There is also some reason for thinking that a triplet excited state can arise from the neutralisation of a positive ion, at least in organic systems:

$$A^+ + e^- \longrightarrow {}^3A^* \qquad (4.14)$$

Excited states having once been formed can return to the ground state by various routes, they can react unimolecularly to give reactive or stable products or they can enter into bimolecular reaction with other molecules. These processes will now be considered.

CONVERSION TO THE GROUND STATE

Fluorescence

An excited molecule can return to the ground state by emitting radiation of a wavelength corresponding to the transition. If excitation had originally been achieved by the absorption of light, then the wavelength of the light emitted on the return to the ground state would be longer than that of the light absorbed. Such a process is known as fluorescence or radiative conversion to the ground state. It can be understood by reference to *Figure 4.1*. It has already been pointed out that the most probable excitation process is represented by the vertical line from the middle of the lowest vibrational level of the ground state. The excited state will thus be formed with excess vibrational energy but this will soon be lost by collision and the excited state will soon achieve its lowest vibrational level. The most probable transition on de-excitation is once again represented by a vertical line (Franck–Condon principle) downwards from the middle of the lowest vibrational level of the excited state. This transition produces a ground state with excess vibrational energy which once again is soon lost by collision. It will be seen from the diagram that the energy quantum in the downward direction is less than that absorbed in the excitation, so that the wavelength of the light emitted is longer than that of the light absorbed. This radiative conversion to the ground state may be represented

$$^1A^* \longrightarrow {}^1A + h\nu \qquad (4.15)$$

Fluorescence usually occurs after a lifetime of about 10^{-9} s of the excited state.

Internal conversion

For a complex molecule there will be a large number of different excited states and the potential energy curves may lie close together. If the curve corresponding to an excited state crosses the ground state curve at a vibrational level lower than that to which the molecule was first excited, it is possible for the molecule to cross from the excited state to the ground state. This process is illustrated in *Figure 4.2*. The molecule is excited from the ground state, *G*, to the excited state, *E*,

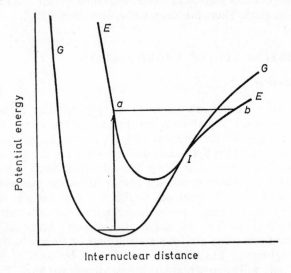

Figure 4.2. Internal conversion

where it would vibrate with an amplitude corresponding to the distance *ab*. Initially the molecule is compressed at the point *a*, and, as it stretches, the energy follows the curve *E* and would eventually reach the point *b*. At *I*, however, the energies and internuclear distances of the ground state and excited state are the same and the molecule may follow the curve *G*. It is thus in the ground state with some excess vibrational energy which it will soon lose by collisions. The molecule has thus been converted to the ground state without any emission of radiation; the process is called internal conversion, provided that the two states have the same multiplicity.

Internal conversion is rather more common between two different excited states of the same multiplicity than between an excited state and the ground state. Thus in polyatomic molecules higher excited states are rapidly converted to the lowest excited state in about 10^{-13} s by internal conversion.

Intersystem crossing and phosphorescence

Although the most probable transition from a singlet ground state is to a singlet excited state rather than to a triplet excited state, the triplet state can be formed if the potential energy curves of the excited singlet and the triplet cross. This is analogous to internal conversion but here the conversion is between two states of different multiplicity and is called intersystem crossing. Although the transition from the triplet excited state to the singlet ground state is optically forbidden, the transition occurs but with a low probability. This means that the lifetime of the triplet state may be relatively long, a normal value being about 10^{-13} s, but it can be as long as several seconds. The triplet state returns to the ground state with the emission of radiation which, owing to the lifetime of the triplet, occurs at a relatively long interval after absorption. Under these circumstances it is known as phosphorescence and the radiation will be of a longer wavelength than that which was initially absorbed. The over-all process may be written

$$^{1}A^{*} \longrightarrow {}^{3}A^{*} \longrightarrow {}^{1}A + h\nu \qquad (4.16)$$

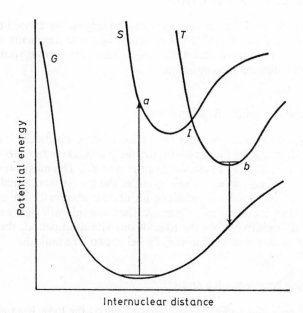

Figure 4.3. Intersystem crossing

and is represented in *Figure 4.3*. The initial excitation is from the ground state G, to the singlet excited state S, at the vibrational level a. The molecule follows the singlet curve and can cross to the triplet curve T at the point I. The triplet state loses vibrational energy until it reaches the zero point vibrational energy of that state at b, when it can emit radiation and return to the ground state.

Non-radiative energy transfer

Electronic excitation energy can be transferred from one molecule to another in a number of ways but each process can be represented by the reaction

$$A^* + B \longrightarrow A + B^* \qquad (4.17)$$

The condition is that the excitation energy of A must be at least equal to that of B. If the electronic energy lost by A in the transfer is greater than that gained by B, the excess is accommodated as vibrational or translational energy of the molecules.

UNIMOLECULAR REACTION

The most usual unimolecular reactions undergone by excited states result in the dissociation of the molecule into two fragments which may be free radicals or stable molecules. There are various routes by which this may be accomplished.

Excitation to a high vibrational level

If the equilibrium internuclear separation of the excited state is much greater than that of the ground state, the potential energy curves for these two states may be drawn as in *Figure 4.4*. The most probable excitation process from the mid-point of the lowest vibrational level of the ground state will produce an excited state with an energy greater than its dissociation energy. The molecule will thus expand until it dissociates within the time of one vibration period, the products of such a dissociation usually being two free radicals.

Excitation to a repulsive state

All the potential energy curves considered so far have been curves corresponding to stable states of a molecule in which there is a

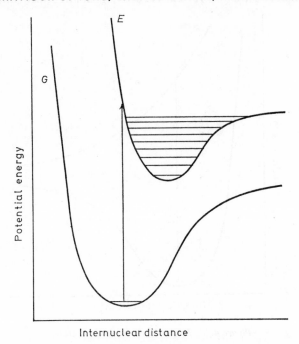

Figure 4.4.   Excitation to a high vibrational level

minimum in the curve at the equilibrium internuclear distance. Some excited states are unstable, having no minimum in the potential energy curve, and are called *repulsive states*. This situation is illustrated in *Figure 4.5*, where the transition from the ground state, *G*, to a repulsive excited state, *R*, is represented. Once again in this case dissociation will occur within the time of one vibrational period.

Predissociation

When a molecule is in an excited state, it may cross to another excited state from which dissociation occurs. Such a process is known as predissociation. It will be appreciated that if the potential energy curve of a stable excited state is crossed by the curve of a repulsive state, there is the possibility of transfer of the molecule from the stable state to the repulsive state and dissociation will ensue.

In the dissociation of excited states as described above, the two

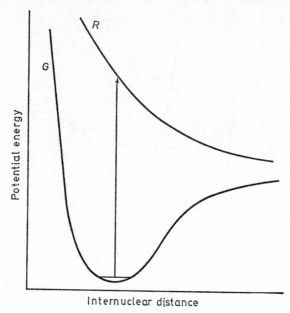

Figure 4.5. Excitation to a repulsive state

electrons forming the bond which is broken are usually divided one to each of the two fragments. In this way the most common products of dissociation are two free radicals as each fragment will have one unpaired electron. The process may be represented

$$A^* \longrightarrow R\cdot + S\cdot \tag{4.18}$$

where $R\cdot$ and $S\cdot$ are free radicals, or if the excited state is an ion,

$$(A^+)^* \longrightarrow R^{\cdot+} + S\cdot \tag{4.19}$$

in which case one of the fragments will be a free radical ion. One of the radicals will probably be in an excited state itself. Such radicals are frequently more reactive than normal free radicals and are called hot radicals.

It will be appreciated that, for the homolytic fission of a chemical bond, the minimum amount of energy required will be the dissociation energy. In the dissociation process, however, it frequently happens that the energy available is greater than the dissociation energy. This can be understood, for example, by reference to *Figure 4.4*. Promotion from the ground state by the most probable transition as indicated results in the energy of the excited state being higher than the dissociation plateau of the curve. The state thus has an energy in excess of the dissociation energy and this excess energy is usually carried off as kinetic energy of the radicals formed by dissociation. Depending on the circumstances of a particular case, then, radicals formed by reactions such as (4.18) or (4.19) may have a variety of kinetic energies. This can be an important consideration in liquid media. Radicals formed with only very small kinetic energies in liquids may not have enough energy to penetrate through the barrier of closely packed molecules of the liquid which surround them. Under these circumstances, the two radical fragments will be 'caged' by the liquid molecules and will recombine, dissipating the dissociation energy as heat. This process is often called the *Franck–Rabinowitch effect* or the *cage effect* and no net chemical change is observed. It follows that chemical changes are more likely to be produced from a dissociation process when the radicals formed have a fair amount of kinetic energy which allows them to escape from the cage. This will require the dissociation process to involve a highly excited state, such as results from charge neutralisation as in reaction (4.3).

In addition to dissociating into free radicals, an excited state may dissociate into stable molecules, although this event is less frequent. It may be represented as

$$A^* \longrightarrow B + C \tag{4.20}$$

Alternatively, if the excited state is an ion, one of the products will be an ion:

$$(A^+)^* \longrightarrow B^+ + C \tag{4.21}$$

Once again, one of the products formed in these processes may be in an excited state.

In addition to the dissociation processes described above, polyatomic excited molecules or ions may undergo unimolecular reactions in a molecular rearrangement process as, for example, the conversion of a *trans*-isomer to a *cis*-isomer.

BIMOLECULAR REACTION

In bimolecular reactions the excited molecule reacts with another molecule and there are four main classes of reaction.

Electron transfer reactions

In these reactions an electron is transferred between the excited molecule and another neutral molecule or an ion. The reaction may be written

$$A^* + B \longrightarrow A^+ + B^- \qquad (4.22)$$

Abstraction reactions

These reactions usually involve the abstraction of hydrogen by the excited molecule from another molecule. As a bond is broken in the process the products are two free radicals:

$$A^* + RH \longrightarrow AH^· + R^· \qquad (4.23)$$

Addition reactions

These reactions are characteristic of triplet excited states rather than singlet excited states. The reactions frequently involve the addition of oxygen, as the ground state of oxygen is, in fact, a triplet state. In general, such reactions may be represented

$$A^* + B \longrightarrow AB \qquad (4.24)$$

Stern–Volmer reactions

These reactions involve the transfer of atoms between an excited state and another molecule which may itself be excited. The reactions may be represented

$$A^* + B \longrightarrow C + D \qquad (4.25)$$

In the foregoing part of this chapter, the reactions of ions and excited states produced by ionising radiations have been described. Both species can give rise to stable products; ion neutralisation reactions can provide excited states and both the ions and excited states can lead to the formation of free radicals. The situation may be summarised as follows:

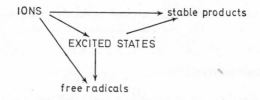

DETECTION AND STUDY OF FREE RADICALS

A free radical is characterised by having an unpaired electron which can render it highly reactive. The reactivity of a radical depends to a large extent on its molecular structure. With some large organic radicals, the odd electron can be distributed over a large volume, thus conferring a great degree of stability on the radical. Smaller radicals and inorganic radicals are usually very reactive. In cases where a radical has two unpaired electrons it may be considered as a diradical. This indicates the importance of the triplet excited state, which has two unpaired electrons and may be regarded, in some respects, as a diradical.

As reactive free radicals can arise from the reactions of ions and excited states produced by ionising radiation, it is necessary to know something of their behaviour for an understanding of radiation chemistry.

One of the most important methods of detecting and identifying free radicals from the point of view of radiation chemistry is *electron paramagnetic resonance spectroscopy* (e.p.r.) or, as it is sometimes called, *electron spin resonance spectroscopy* (e.s.r.). An unpaired electron, such as is present in a free radical, has a magnetic moment due to its spin. When the free radical is placed in a magnetic field, the unpaired electron orientates itself with respect to the field so that its magnetic moment is either with the external field or against it, depending on the direction of spin of the electron. In the absence of a magnetic field each free radical in a collection of radicals of the same species will have the same energy. In a magnetic field those radicals whose unpaired electrons are aligned with the field will have a lower energy and those with unpaired electrons orientated against the field will have a higher energy. The external field thus causes the splitting of an energy level. The unpaired electrons can be promoted from the lower to the upper state by the absorption of radiation of quantum energy corresponding to the difference in the split energy levels. This energy difference will depend upon the magnetic flux density and, with a flux density of about 3500 oersted (0·35 tesla) the frequency of the radiation required to promote an electron is about 10^4 MHz and the corresponding wavelength is about 3 cm. The electrons are thus promoted by the absorption of energy, which causes their orientation to be reversed and with appropriate equipment a spectroscopic absorption line can be observed. Normally the frequency of the electromagnetic radiation is fixed and the magnetic flux density is varied, when a plot is obtained of

F

energy absorption against flux density. A typical plot is shown in *Figure 4.6.*

In addition to the splitting of the energy level due to the external magnetic field, further splitting can arise owing to the presence, in the

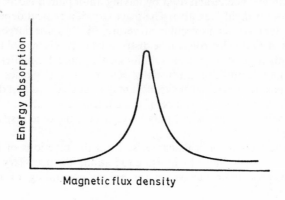

Figure 4.6. Electron paramagnetic resonance absorption peak

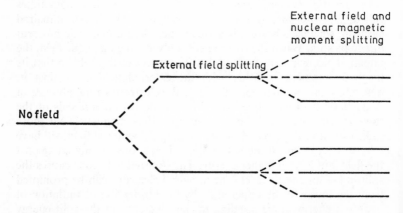

Figure 4.7. Splitting of energy levels by magnetic fields

free radical, of a neighbouring atomic nucleus with a nuclear magnetic moment. Such additional splitting is represented diagrammatically in *Figure 4.7*, and the extra transitions which are possible give rise to further absorption peaks in the spectrum. The number of extra peaks will depend on the spin of the nucleus causing the extra splitting. It is thus possible to detect a free radical by its e.p.r.

spectrum and it may also be identified by the fine structure of the spectrum.

The technique of e.p.r. can be used for the examination of stable free radicals and of reactive radicals which have been frozen in a solid. In this latter case, the mobility of the radicals and hence the possibility of their reaction is reduced. It is thus possible to study radicals formed by the irradiation of solids by this technique.

In reactions involving free radicals as intermediates, these reactive entities are present at any time only in extremely low concentrations. This is due to the fact that they are usually very reactive and thus have only very short lifetimes. Under these circumstances the free radicals are not present in high enough concentrations to be detected directly and their existence is inferred from the nature of the reaction products. It is possible, however, to generate free radicals in concentrations sufficiently large to be observed spectroscopically by irradiating a system with an intense flash of light. This technique is known as *flash photolysis* and has provided a great deal of information on free radical intermediates in photochemical and other reactions.

Of more direct interest to radiation chemistry is the technique of *pulse radiolysis*, which is the analogue of flash photolysis in radiation chemistry and was first used in about 1960. A pulse of X-rays or fast electrons is used in place of the light flash. In this way the radicals and solvated electrons generated by ionising radiations can be directly observed. A high concentration of these reactive species is produced by the high intensity of the pulse and the absorption spectrum of the system is observed immediately afterwards. The presence of radicals and solvated electrons is inferred from the spectrum, and if we follow the change in the absorbance with time, the kinetics of the reactions undergone by the transient species can be directly observed.

The shapes of the radiation vessels must be arranged to accommodate the volume of solution which can be uniformly irradiated by the pulse. 10 MeV electrons can penetrate about 5 cm in water and arrangements can be made for a pulse of 10 MeV electrons to irradiate uniformly a vessel containing about 5–10 cm^3 of liquid.

The absorption spectrum of the system is observed at various intervals after the occurrence of the initiating pulse by arranging for a spectroscopic lamp to flash by being triggered automatically through an electronic delay circuit, which is operated by the initiating pulse. The optical path length through the system can be effectively increased by multiple reflections or by total internal reflection using a cell in the form of a spiral tube. The light emerging from the system passes through a monochromator and is subsequently detected by a photomultiplier, the response of which is displayed on an oscilloscope and may be photographed. The time base of the oscilloscope is triggered by a sensor which reacts to the initiating pulse of radiation and, to resolve the absorption due to the very short-lived species, the oscilloscope must have a microsecond or even a nanosecond time base.

The fastest reaction which can be studied by this technique is determined by the duration of the initiating pulse. The duration of a pulse is typically 10^{-6}–10^{-9}s, so that reactions with half-lives of 10^{-9}s can be studied. This means that the reactions of radicals and solvated electrons can be observed and it also allows the study of the reactions of some excited states.

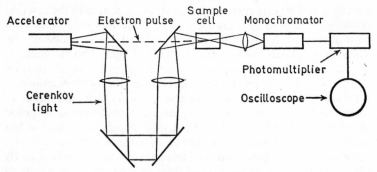

Figure 4.8.   Stroboscopic pulse radiolysis

The most useful source of ionising radiation for pulse radiolysis is the linear accelerator, although Van de Graaff generators with an electron gun and accelerating tube may also be used. Ion accelerators do not have the right characteristics in terms of energy per pulse, time of pulse and volume distribution of energy deposition.

A recent development of pulse radiolysis is *stroboscopic pulse radiolysis*, which decreases the time scale from 1 ns to about 20 ps. This technique depends on two facts. First, a 40 ns pulse of electrons from a linear accelerator has a fine structure and may contain about 100 fine structure pulses about 0·02 ns wide with a separation of about 0·35 ns. Second, in addition to the modes of energy loss of charged particles mentioned in Chapter 3, fast electrons lose a small amount of energy as *Cerenkov radiation*, which takes the form of a blue glow. In stroboscopic pulse radiolysis the absorbing species formed by one of the fine structure pulses are analysed by the light of the Cerenkov radiation from the preceding fine structure pulse. The experimental arrangement is illustrated in *Figure 4.8*. The electron beam from the accelerator passes through an air space before penetrating two thin mirrors and passing on to the sample cell. The Cerenkov radiation from a fine structure pulse creates a light flash lasting less than 0·02 ns in the air space before the thin mirror. This light is diverted by the mirrors, so that there is a delay of about 0·3 ns before it reaches the sample cell, where it arrives in time to analyse the transient species formed in the cell by the next fine structure electron pulse. By adjusting the path length of the Cerenkov light, the variation in the absorbance of the transient species may be followed for about 0·3 ns between successive fine structure pulses.

This technique of picosecond pulse radiolysis may be useful in studying the reactions of 'dry' electrons before they become solvated.

REACTIONS OF FREE RADICALS

The free radicals which are formed as a result of the reactions of ions and excited species can, in common with radicals formed by other methods, react in a variety of ways, either unimolecularly or bimolecularly, with other radicals or with stable molecules.

UNIMOLECULAR REACTIONS

Rearrangement

It has been pointed out earlier that the stability or, conversely, the reactivity of a radical depends to some extent upon its structure. In some cases a radical can, by molecular rearrangement, achieve a more stable structure. Such reactions may be represented

$$AB\cdot \longrightarrow BA\cdot \qquad (4.26)$$

where the migration of a group or atom within the radical causes a change in position of the unpaired electron.

Dissociation

Large organic radicals can sometimes dissociate into a smaller radical and a stable molecule if such a process is energetically favourable:

$$AB\cdot \longrightarrow A\cdot + B \qquad (4.27)$$

In this case the stable molecule, B, is usually an unsaturated compound.

In addition to this type of dissociation, if the radical is formed in an excited state it may undergo a dissociation reaction as has already been discussed for excited states.

BIMOLECULAR REACTIONS

Combination

Two free radicals can combine together sharing their unpaired electrons to form a bond:

$$R_1^{\cdot} + R_2^{\cdot} \longrightarrow R_1R_2 \tag{4.28}$$

The energy liberated in this process is equal to the dissociation energy of the bond formed. If the resultant molecule is large, the energy can be distributed over the molecule and subsequently be lost by collisions. If the resultant molecule is small, it may break up again into radicals unless a third body is present to remove the energy. In the liquid phase there is no shortage of third bodies, but in the gas phase most recombination reactions occur on the walls of the containing vessel, the wall acting as a third body.

Disproportionation

Two radicals can react together to give stable products by the transfer of an atom from one radical to the other. In most cases the atom transferred is hydrogen and the reaction often occurs with organic radicals when one of the products is saturated and the other unsaturated:

$$RH^{\cdot} + RH^{\cdot} \longrightarrow RH_2 + R \tag{4.29}$$

Addition

A common reaction of radicals is addition to an unsaturated compound:

$$R^{\cdot} + {>}C{=}C{<} \longleftrightarrow {>}CR{-}\overset{\cdot}{C}{<} \tag{4.30}$$

In these reactions, of course, the radical is not destroyed, but merely forms a larger radical which can continue to undergo radical reactions.

Abstraction

Radicals can abstract atoms from organic compounds to form a stable molecule and another radical, the stability of the radical so formed usually being greater than that of the original radical. The atom which is abstracted is usually a halogen or a hydrogen atom; and if we represent these by X, the abstraction reaction may be written

$$R\cdot + AX \longrightarrow RX + A\cdot \qquad (4.31)$$

Electron transfer

Radicals can react with ions in electron transfer processes. An electron is usually transferred to the radical to form a negative ion and the original ion may be oxidised to another ion or a free radical:

$$\left.\begin{aligned}R\cdot + A^+ &\longrightarrow R^- + A^{2+}\\ R\cdot + B^- &\longrightarrow R^- + B\cdot\end{aligned}\right\} \qquad (4.32)$$

In the second case the new radical will, of course, undergo further reaction.

Reaction with oxygen

Radicals react readily with oxygen to form peroxy radicals and these reactions are most important in radiation chemistry. The ground state of oxygen is a triplet state, having two unpaired electrons, and thus behaves as a diradical. The oxygen molecule adds on to a radical, one of the oxygen's unpaired electrons being shared with the radical's unpaired electron to form a bond. There remains the other unpaired electron from the oxygen molecule which confers the properties of a radical on the addition product:

$$R\cdot + O_2 \longrightarrow R\dot{O_2} \qquad (4.33)$$

These reactions can be important in increasing oxidation yields in radiation chemical reactions.

RADIATION CHEMICAL YIELDS

It has been seen that the effect of the transfer of energy to a medium by ionising radiation is, in the first instance, the production of reactive entities such as ions, excited states and free radicals. These species can react with themselves or with substrates which may be present in the system and eventually give rise to stable products. For given experimental conditions, the amount of product or the *yield* resulting from the action of the radiation will depend on the amount of energy deposited in the system. Yields in radiation chemistry have been expressed in two ways.

IONIC YIELD

In early studies of radiation chemistry the production of ions in gaseous systems could be readily observed and the ions were thought to exert a controlling influence on the chemical changes produced. Yields were thus expressed as ionic yields, usually denoted by M/N, where M is the number of molecules which have undergone chemical change and N is the number of ions produced in the system. The ionic yield is thus the number of molecules changed per ion produced. The energy required to produce an ion pair in gases can be measured, so that the ionic yield is an implicit statement of the yield of product per amount of energy absorbed. In fact, most ionic yields have been expressed as number of molecules changed per 32·5 eV absorbed, this being the average amount of energy dissipated for the formation of an ion pair in air.

THE *G*-VALUE

In condensed phases, the ionic yield has less significance as the number of ions formed can not be determined and has to be calculated by the assumption of a value for the energy dissipated for the formation of an ion pair in the condensed phase. In the past this has been assumed (for lack of better information) to be the same as for air. In principle, the amount of energy absorbed by a condensed medium from ionising radiation can be measured directly and yields are expressed in terms of a G-value, which is the number of molecules changed per 100 eV of energy absorbed. In this way no particular significance is attributed to the ions produced.

If the absorption of 100 eV of energy by a system resulted in the formation of x molecules of a substance A, this data would be expressed as $G(A) = x$. In some experiments, the destruction of a particular compound is of interest. If, for example, y molecules of substance B were destroyed by the absorption of 100 eV of energy, the result may be expressed $G(-B) = y$, where the minus sign indicates a loss. Sometimes subscripts are added to specify the conditions more closely, such as the type of radiation used, the presence or absence of oxygen, etc.

Most radiation chemical yields are now expressed as G-values, even for gaseous systems, as the ionic yield may tend to imply a controlling influence exerted by the number of ions formed on the amount of chemical reaction. It is now realised that excited states may play an important part in radiation chemistry and it is best to relate the over-all chemical change to the dose of energy absorbed.

Obviously, in order to express the results in G-values, the dose of energy absorbed by a system under particular conditions must be known. The determination of the amount of energy absorbed is known as *dosimetry* and forms the subject of the next chapter.

5
Dosimetry

To determine a radiolytic yield accurately, it is of importance to measure the energy uptake precisely. The most direct method for doing this is a calorimetric method, in which the increase in temperature of a block of inert material exposed to the radiation is measured. There are many experimental difficulties with this method and it is therefore inconvenient for routine purposes. The two most widely used techniques—ionisation dosimetry and chemical dosimetry—are both indirect. Several other methods of measuring the dose of energy absorbed by a system will also be discussed in this chapter.

IONISATION DOSIMETRY

One of the first observations made with ionising radiations was their ability to discharge an electroscope. This occurred as a result of the formation of positive ions and electrons in the medium through which the radiation passed. Very soon more subtle instruments for detecting and measuring radiation were constructed, such as ionisation chambers and condenser-r-chambers (*Figure 5.1*). These, again, consist of air-filled chambers containing two electrodes across which a potential difference of about 100–200 V is applied. When the chamber is exposed to radiation, the electrons and positive ions are collected on the anode and the cathode respectively.

The first unit to describe a quantity of radiation was the *roentgen*, r, which was defined in terms of electrical discharge as follows. '1 roentgen is that quantity of x- or γ-radiation whose secondary corpuscular emission produces in 1 cm³ of air at S.T.P. 3.3×10^{-10} C of electricity of either sign.' The instruments mentioned above may thus be used to measure, in terms of roentgens, the dose of radiation to which they are exposed.

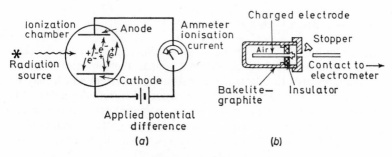

Figure 5.1. (a) Simplified version of ionisation chamber; (b) condenser-r-chamber

CONDENSER-R-METERS

In the condenser-r-meter the electrodes are initially charged and the collection of the ions and electrons generated by the radiation causes a change in the potential difference between the electrodes, which can be measured directly with an electrometer. The instrument is simply a condenser which is partially discharged by the ions and electrons. The relationship between the potential difference, U, between the plates of a condenser, the charge it holds, Q, and its capacitance, C, is given by

$$Q = CU \qquad (5.1)$$

Suppose that a condenser-r-meter of capacitance C is initially charged with a quantity of electricity Q_1, the resulting potential difference between the electrodes being U_1. Suppose further that after exposure of the instrument to radiation the charge is reduced to Q_2, the corresponding potential difference being U_2. Then the amount of charge lost as a result of exposure $(Q_1 - Q_2)$ is given by

$$(Q_1 - Q_2) = (U_1 - U_2)C \qquad (5.2)$$

U_1 and U_2 may be measured with an electrometer; and if the capacitance of the condenser-r-meter is known, $(Q_1 - Q_2)$ may be calculated. If the volume of the chamber is known together with the temperature and pressure of air inside the chamber, the volume of air inside the chamber can be expressed as a volume at S.T.P. If the charge generated by the radiation in this volume of air at S.T.P. is known, the dose can readily be expressed in roentgens from the definition of this unit above.

IONISATION CHAMBERS

In the ionisation chamber the potential difference between the electrodes is maintained by an external source, the ions and electrons moving towards the electrodes constituting a current which may be measured by an external circuit. The observed current depends on the radiation dose rate and on the applied potential between the plates. *Figure 5.2* shows the relationship between current and applied potential for a fixed dose rate. When no potential difference is applied,

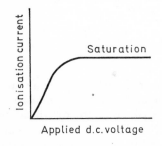

Figure 5.2. Variation of current with applied potential in an ionisation chamber

only a few ion pairs are collected on the plates. The others recombine in the gas phase or on the other surfaces of the chamber. As the applied voltage is increased, more and more ion pairs are separated and collected on the electrodes and the current flowing between the electrodes increases. When the applied voltage is such that all the possible ions and electrons are being collected, the current reaches a limiting value, known as the saturation current. This saturation current is a measure of the total number of ion pairs formed per unit time between the electrodes. This is, of course, the rate of formation of the ions which is a measure of the *dose rate* to which the chamber is exposed.

As the roentgen is defined in terms of 1 cm³ of air, the perfect ionisation chamber should have walls made of air. Walls made of any other material would have a different absorption and so change the equal distribution of secondary electrons (secondary corpuscular emission) entering and leaving the air volume of the chamber. Walls denser than air cause more electrons to enter the air space than leave it. This would result in collecting a greater charge than that produced by absorption of the ionising radiation in air. After meticulous experimentation, air ionisation chambers were constructed which absolutely follow the definition of the roentgen and so gave an absolute measurement (*Figure 5.3*).

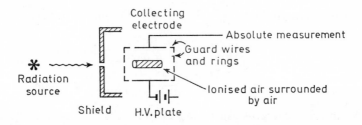

Figure 5.3. Absolute ionisation chamber

For most routine cases, however, it is quite impossible to work with such a refined chamber and an 'air-equivalent' instrument is then employed. This has walls made of a plastic such as Bakelite, which has an electron density similar to that of air. With this technique chambers can be made which are quite small with volumes of only a few cubic centimetres. These small chambers are known as 'thimble' ionisation chambers and are very useful for plotting isodose curves. Also, as will be seen below, they are suitable for measuring the actual energy absorbed in a medium.

EXPOSURE DOSE AND ABSORBED DOSE

It will be noted that the roentgen is defined through an electrical discharge in air. It is not a measure of the energy absorbed in the chamber. Thus the roentgen is a measure of exposure rather than of energy uptake. If a system in a radiation beam is replaced by an ionisation chamber, the reading in roentgens will be a measure of the amount of radiation to which the system was exposed. It will not give any information about the energy absorbed by that system from the radiations. Nevertheless, it is possible as will be seen below, with the use of some additional information, to convert such a reading into the energy absorbed in air and in the system.

One piece of information required to convert exposure dose to absorbed dose in air is the amount of energy required to form an ion pair in air. With the use of α-particles, which could be completely contained in the volume of an ionisation chamber owing to their short range, it has been found that the energy dissipated for the formation of an ion pair in air is 34 eV. Now a roentgen is defined as producing $3\cdot33 \times 10^{-10}$ C of charge per cubic centimetre of air at S.T.P. The charge on an electron or monovalent ion is $1\cdot6 \times 10^{-19}$ C and the density of air at S.T.P. is $0\cdot001293$ g cm^{-3}. Thus

$$1 \text{ roentgen produces } \frac{3\cdot33 \times 10^{-10}}{1\cdot6 \times 10^{-19}} \text{ ion pairs/cm}^3 \text{ air at S.T.P.}$$

$$= \frac{3\cdot33 \times 10^{-10}}{1\cdot6 \times 10^{-19} \times 0\cdot001293} \text{ ion pairs/g air}$$

$$= \frac{3\cdot33 \times 10^{-10} \times 5\cdot44 \times 10^{-18}}{1\cdot6 \times 10^{-19} \times 0\cdot001293} \text{ J/g air}$$

As absorbed energy is usually expressed in rads, the definition being 1 rad $= 10^{-5}$ J g^{-1}, the exposure of air to 1 roentgen of radiation leads to the absorption of

$$\frac{3\cdot33 \times 10^{-10} \times 5\cdot44 \times 10^{-18}}{1\cdot6 \times 10^{-19} \times 0\cdot001293 \times 10^{-5}} \text{ rad in air}$$

$$= 0\cdot87_7 \text{ rad in air}$$

It can therefore be said that the exposure of air to 1 roentgen of radiation leads to the absorption of $0\cdot87$ rad of energy. Exposure of other materials to 1 roentgen will not result in the same energy absorption. This is because they have different mass absorption coefficients (μ/ρ) and, in respect of the secondary corpuscular emission, different mass stopping powers $\left(-\frac{1}{\rho}\frac{dE}{dx}\right)$. When these two factors are taken into account relative to air, then the energy absorbed by any system can be found:

Energy absorbed by medium = Energy absorbed by air

$$\times \frac{\text{mass absorption coefficient in medium}}{\text{mass absorption coefficient in wall of chamber}}$$

$$\times \frac{\text{mass stopping power in wall of chamber}}{\text{mass stopping power in air}} \qquad (5.3)$$

or

$$E_{\text{medium}} = E_{\text{air}} \times \frac{(\mu/\rho)_{\text{medium}}}{(\mu/\rho)_{\text{wall}}} \times \frac{\left(-\frac{1}{\rho}\frac{dE}{dx}\right)_{\text{wall}}}{\left(-\frac{1}{\rho}\frac{dE}{dx}\right)_{\text{air}}} \qquad (5.4)$$

In the particular case of an air-equivalent ionisation chamber where the wall has the same electron density as air, the relative mass stopping term becomes unity and the equation reduces to

$$E_{medium} = E_{air} \times \frac{(\mu/\rho)_{medium}}{(\mu/\rho)_{air}} \qquad (5.5)$$

As was shown in Chapter 3, the mass absorption coefficient of a material varies with the energy of the incident photon. Mass absorption coefficients of materials with similar atomic numbers vary in the same way with changes in the incident photon energy, so that little variation is observed in the ratio of these mass absorption coefficients between incident photon energies of 0·1 and 1·5 MeV. Air, water and tissue have similar atomic numbers and the ratio between the mass absorption coefficients of water or tissue and air is approximately 1·11 over the incident energy range quoted. In this case exposure of water or tissue to 1 roentgen of radiation results in an energy absorption of 0·97 rad. At energies below 0·1 MeV the ratio decreases but does not become less than 1·00.

For materials with atomic numbers very different from that of air, considerable variations develop in the ratio of the mass absorption coefficients with changes in the incident photon energy. As can be seen from *Table 5.1*, these differences are particularly noticeable in the low-energy region (0·1 MeV; photoelectric effect) and also at very high energies (10 MeV; pair production).

It is evident from *Table 5.1* that when an ionisation chamber is used to determine the absorbed dose, the energy of the incident ray must be precisely known. From this the correct mass absorption coefficient can be found and the relative mass absorption ratio of the system to air may be calculated. The absorbed dose can thus be obtained from equation (5.5). Tables of mass absorption coefficients for different materials at various incident energies can be obtained from the literature. In all this, it will be remembered that the walls of the chamber are air-equivalent.

The absorbed dose can also be measure *in situ* by inserting a specially sealed thimble ionisation chamber into the medium under investigation. If the volume of the chamber is very small compared with that of the medium, the electrons entering the cavity all come from the absorption of photons in the medium. No photon absorption takes place in the chamber and its thin air-equivalent walls make no contribution. The secondary electron flux across the chamber is the same as that in the medium and it is this which is measured by the chamber. Any measuring system conforming to these conditions is said to obey the *Bragg–Gray cavity principle*. Because the chamber

Table 5.1 Variation of absorbed energy with incident photon energy

Energy region	Incident photon energy/ MeV	Material	Absorbed energy/rad r^{-1}
Photoelectric effect	0·01	bone	5·0
		water	0·8
		fat	0·4
	0·1	bone	1·4
		water	0·9
		fat	1·0
Compton effect	1·0	bone	0·7
		water	1·0
		fat	1·1
	10·0	bone	0·7
		water	1·0
		fat	1·1
Pair production	100·0	bone	1·3
		water	0·8
		fat	0·7

measures only electrons arising in the medium, no ratios of absorption coefficients are involved and therefore the dose absorbed by the medium will be given by

$$(\text{dose absorbed})_{\text{medium}} = (\text{dose absorbed})_{\text{air}} \times \frac{(\text{mass stopping power})_{\text{medium}}}{(\text{mass stopping power})_{\text{air}}}$$

$$= 0·87 \times \frac{\left(-\frac{1}{\rho}\frac{dE}{dx}\right)_{\text{medium}}}{\left(-\frac{1}{\rho}\frac{dE}{dx}\right)_{\text{air}}} \quad (5.6)$$

The mass stopping power ratio of water to air works out to have a value of 1·15 for cobalt-60 γ-rays or 1·0 MeV photons. Thus it can be calculated that with such γ-rays the dose absorbed in water per roentgen of exposure dose is 1·01$_9$ rad. It should be noted, however, that with the thimble ionisation chamber located in the medium the exposure dose which it measures will not be the same as the exposure dose at the same point if the thimble were surrounded by air. The medium between the ionisation chamber and the source will affect the quality of the radiation reaching the chamber. It may thus be advisable to indicate whether the exposure dose was measured as an *air-dose* or a *cavity-dose*.

CHEMICAL DOSIMETRY

Chemical dosimetry is the measurement of absorbed dose by chemical means. For radiation chemists it is by far the most convenient and popular method of measuring absorbed dose. For use as a dosimeter, a chemical system must have an accurately known radiolytic yield which has previously been determined by a calorimetric method or by ionisation dosimetry. In addition, it is an advantage for the reactant or the products of irradiation to be analysed easily. If a system has a known G-value of, say, x molecules of a particular product per 100 eV of energy absorbed, then exposure of the system to radiation followed by an analysis of the product will give the amount of energy absorbed by the medium under the conditions employed. Some chemical dosimeters will be discussed below.

THE FRICKE DOSIMETER

The best-known chemical dosimeter is the Fricke dosimeter, which is based on the oxidation of ferrous ions to ferric ions as a result of irradiation. The system consists of a 10^{-3} mol dm^{-3} solution of ferrous ammonium sulphate $[FeSO_4.(NH_4)_2SO_4]$ in 0·4 mol dm^{-3} H_2SO_4, the solution being made up with triply distilled water. All the reagents should be of A.R. quality and the glassware (i.e. radiation vessels) should be washed out with chromic acid and rinsed with triply distilled water. The latter is prepared by redistilling distilled water successively from alkaline potassium permanganate and then acid potassium dichromate solutions. These precautions are necessary to avoid the presence of organic impurities, which would increase the yield of ferric ions in the dosimeter. Addition of a little sodium chloride (10^{-3} mol dm^{-3} NaCl) to the system prevents the formation of organic free radicals and allows the use of triply distilled water to be relaxed a little.

When such a solution is exposed to ionising radiations, some of the ferrous ions will be oxidised to ferric ions and the radiolytic yield of ferric ion, $G(Fe^{3+})$, depends upon several factors. First, the presence or absence of oxygen affects $G(Fe^{3+})$ but the Fricke dosimeter employs air-saturated solutions, so that the quantity of interest is the radiolytic yield of ferric ion in the presence of air, which will be denoted by $G(Fe^{3+})_{air}$. The yield also depends upon the LET and the energy of the radiation used. The variation of $G(Fe^{3+})_{air}$ with the type of radiation is given in *Table 5.2*.

The reason for the variation of $G(Fe^{3+})_{air}$ with LET is due to a

G

variation of free radical yield (see Chapter 6). For the purposes of dosimetry, all that is required is a knowledge of the type and, if possible, the energy of the radiation employed. It will be noted that the absolute method by which $G(Fe^{3+})_{air}$ has been related to a calorimetric measurement was carried out by Hochanadel and Ghormley in 1953. They found that 15·6 ions of ferrous were oxidised to ferric for every 100 eV of energy absorbed from a cobalt-60 source.

Table 5.2 Variation of $G(Fe^{3+})_{air}$ with type of radiation

Type of radiation	LET/ eV nm⁻¹	$G(Fe^{3+})_{air}$
Cobalt-60 γ-rays	0·2	15·6
X-rays (100 kV, filtered)	0·92	14·7
Phosphorus-32 β-particles	–	15·2
Accelerated deuterons	5·5	11·0
Helium ions	20	8·0
Polonium-210 α-particles	90	5·1
Uranium-235 fission fragment	–	c. 3

$G(Fe^{3+})_{air}$ is independent of dose rate up to rates of 10^8 rad s⁻¹, but at dose rates greater than about $1·7 \times 10^2$ rad s⁻¹ care has to be taken because all the oxygen dissolved in the solution may be consumed. If this occurred, the yield of ferric would drop, which in turn would cause an error in the dosimetry. It is therefore of great importance to make sure that the solution remains fully saturated with air (or oxygen) throughout the irradiation.

Once the type of radiation to be employed, and thus $G(Fe^{3+})_{air}$, is known and full aeration of the solution has been ensured, the absorbed dose can be found. The ferrous solution is placed in the irradiation vessel and exposed to the ionising radiation, and the amount of ferrous changed to ferric is measured. If $G(Fe^{3+})_{air}$ is known, the energy absorbed by the solution in the vessel in the time exposed can easily be calculated.

There are two main methods for determining the amount of ferrous ion which is oxidised to ferric ion as a result of irradiation. In one method the amount of ferric ion produced is determined by measuring the absorbance of the solution spectrophotometrically at a wavelength of 305 nm, where it is known that the ferric species in solution has a molar extinction coefficient of 2193 at 25° C. As the molar extinction coefficient changes by about 0·7 % per kelvin, the necessary corrections have to be made to its value if the absorbance of the solution is measured at some temperature other than 25° C. The other main method involves the titration of the remaining ferrous ions with standard ceric ammonium sulphate solution using 2,2'-bipyridyl as an indicator.

By either of these analytical methods, it is possible to arrive at the number of ferric ions formed per cm^3 of solution from which the dose of energy absorbed by the solution may be calculated. Suppose that γ-rays are being used for which $G(Fe^{3+})_{air} = 15.6$ molecules per 100 eV and that x molecules of ferric ion are formed per cubic centimetre of solution. The dose is then given by

$$\text{dose} = \frac{100x}{15.6} \text{ eV cm}^{-3}$$

or, since 1 eV $= 0.16$ aJ,

$$\text{dose} = \frac{16x}{15.6} \text{ aJ cm}^{-3} \tag{5.7}$$

If the dose is to be expressed in rads, it is necessary to know the density of the dosimeter solution, as 1 rad $= 10^{-5}$ J g^{-1}. For the Fricke dosimeter the density between 15° and 25 °C is 1.024 g cm^{-3} and thus equation (5.7) may be written as

$$\text{dose} = \frac{16x \times 10^{-13}}{15.6 \times 1.024} \text{ rad} \tag{5.7a}$$

In this way, the total dose of absorbed energy or the dose rate can easily be determined.

If a system with a density other than that of the dosimeter solution is to be subsequently studied, it must be remembered that the absorption of energy depends upon the density of the medium and allowances for this effect are made on a proportional basis. If the system to be studied is water or a dilute aqueous solution of density 1 g cm^{-3}, it will absorb less energy owing to its lower density. In this particular case the dose as measured by the dosimeter and given by equation (5.7a) must be multiplied by the factor $1/1.024$, in order to arrive at the dose which will be absorbed by the system under investigation.

The volume, position with respect to the radiation source and the shape of the system should be the same as that of the dosimeter. Volume and position present little difficulty, but shape replication may require the preparation of a special phantom.

SOME ALTERNATIVE CHEMICAL DOSIMETERS

When the dose rate to the system under investigation is likely to be so high that the dissolved oxygen in the Fricke dosimeter is soon exhausted, other chemical dosimeters can be employed which do not depend on the presence of oxygen. One of these is the ceric sulphate dosimeter, in which ceric ions are reduced to cerous ions as a result of energy absorption.

The dosimeter consists of ceric sulphate dissolved in sulphuric acid solution of concentration 0·4 mol dm^{-3} H$_2$SO$_4$, the solution being made up with triply distilled water. The concentration of ceric sulphate is usually in the range 10^{-4} to 10^{-1} mol dm^{-3} Ce(SO$_4$)$_2$, so that approximately 50 % of the ceric ions are reduced in a reasonable time with the dose rates usually encountered. The amount of change produced by irradiation is obtained by determining the amount of ceric ion remaining in the solution. With the more dilute ceric sulphate solutions this is done spectrophotometrically by measuring the absorbance of the solution at a wavelength of 320 nm. The more concentrated solutions are usually analysed by titration with standard ferrous ammonium sulphate solution. The ceric dosimeter can be standardised with respect to energy absorption from ionising radiation against a Fricke dosimeter. With cobalt-60 γ-rays and concentrations of Ce(SO$_4$)$_2$ between 10^{-5} and 5 $\times$ 10^{-2} mol dm^{-3}, $G(\text{Ce}^{3+})$ = 2·45 molecules per 100 eV. As this G-value is much less than that for the Fricke dosimeter the variation of $G(\text{Ce}^{3+})$ with LET will be proportionately less than that of $G(\text{Fe}^{3+})$. In general, the best way of dealing with any variation in the quality of the source and the nature of the dosimetric system is to take $G(\text{Fe}^{3+})_{\text{air}}$ = 15·6 molecules/100 eV for cobalt-60 γ-rays as the standard and then calibrate the G(dosimeter) value for the particular radiation from this.

An alternative to the ceric dosimeter which is also independent of oxygen is achieved by using a Fricke dosimeter with the addition of an excess (10^{-2} mol dm^{-3}) of cupric ions to the solution. The radiolytic yield of ferric ions in this case, is given by $G(\text{Fe}^{3+})_{\text{Cu}^{2+}}$ = 0·9 molecules/100 eV in the presence or absence of oxygen. Again, if it is ensured that the standardising Fricke dosimeter solution is used under the correct conditions (plenty of oxygen), this system can be calibrated initially against $G(\text{Fe}^{3+})_{\text{air}}$ with cobalt-60 γ-radiation.

In pulse radiolysis the dose rates are often very high and the Fricke dosimeter may be unsuitable, particularly where absorbance due to ferric sulphate is measured a few microseconds after the pulse. Not all the radicals or molecular products will have had time to react with ferrous ions by then. Under these circumstances, it is often more convenient to determine the dose per pulse by measuring the absorbance at 720 nm due to the hydrated electron in triply distilled water. Another species which is convenient for pulse radiolysis dosimetry is (SCN)$_2^-$. This is formed with a G-value of 2·9 molecules/100 eV, by pulsing an aerated potassium thiocyanate (1 to 5 $\times$ 10^{-2} mol dm^{-3} KSCN) solution. The (SCN)$_2^-$ radical can be observed by its absorbance at 500 nm a few microsceonds after the pulse.

A number of solid dosimeters have been developed for high doses. Amongst these are dyed Cellophane, polyvinyl chloride and red Perspex for package irradiation plants where the doses are in the Mrad range. Other solid dosimeters which have been tried are the silver activated phosphate and cobalt borosilicate glasses. In all cases a darkening is observed as a result of energy absorption from radiation. The degree of darkening, as measured with a photometer, can be related to the absorbed dose. The difficulty with these dosimeters is that the discoloration fades with time and increasing temperature. It is necessary to take account of these parameters both during the calibration and later when the solid is used as a dosimeter.

Thermoluminescent dosimeters, such as calcium and lithium fluoride (doped with a little manganese), have been developed with respect to personnel and general dosimetry. Exposure of these substances to ionising radiations results in the incorporation of defects which can be removed at a later time by heating. As the defects disappear, light is emitted, the intensity of which is directly proportional to the radiation absorbed. The virtue of this dosimeter is that it covers such a wide range.

Gels incorporating radiation-sensitive dyes (e.g. methylene blue) have been used as dosimeters. Their interest lies in the fact that the change of absorbed dose with depth of the gel can be directly observed.

Often in industrial radiation, just a quick colour indicator is needed on the side of a container to show whether the packet has been sufficiently irradiated. Suitable systems for this purpose are organic halogen compounds (e.g. chloroform) which decompose on the absorption of radiation to give hydrochloric acid. The whole system can be incorporated with a pH paper to give the desired colour change irradiation indicator. This type of system using a pH indicator can also be used in gels.

MEASUREMENT OF LOW DOSES

Counting equipment showing the arrival of individual rays or particles is used for measuring low doses. These are of interest to radiation chemists, biologists and health physicists. By this technique, each individual photon or particle absorbed in the detector is recorded. The standard detectors consist either of a Geiger–Müller tube or of a scintillation head with a photomultiplier. In both detectors the ionising species are converted into an electronic pulse, which is amplified and counted. The count per unit time is a measure of the radiation flux. If the pulses are fed into a CR circuit, the current in the circuit or the voltage drop across the resistance will give a direct reading of counts per second on a meter. It is possible to recalibrate this rate meter into a dose rate meter by means of a calibrated radioactive source. Thus, for γ-rays, the dose rate at a given distance from a point source may be calculated from the equation

$$\frac{\text{dose rate}}{\text{(rad/h)}} = \frac{\text{specific } \gamma\text{-ray constant} \times \text{source strength (curies)}}{[\text{distance of detector from source (metres)}]^2}$$

Values of specific γ-ray constants which depend on the energy of the radiation are given in *Table 5.3* for a variety of sources.

Table 5.3 Specific γ-ray constants

Radioactive isotope	Specific γ-ray constant
Cobalt-60	1·32
Caesium-137	0·33
Radium-226	0·83
Iodine-131	0·22
Sodium-24	1·84

Other equations are available to calculate dose rates from other sources, geometries and detector situations.

The sources used for calibration of health physics instruments are small (in the millicurie region), because the dose rates which these instruments are expected to read should not be far from background radiation. In health physics a new unit is used because equal absorbed radiation doses do not necessarily give rise to the same biological damage. To estimate the radiation hazard with respect to man, the absorbed dose in rads must be multiplied by one or more weighting factors to obtain the true dose equivalent. When radiations of different LET are compared, the absorbed dose should be multiplied by a quality factor (QF) which corrects for biological LET effects. The dose-equivalent unit is the *rem* and can be defined as

$$1 \text{ rem} = 1 \text{ rad} \times QF$$

With LET values of 3·5 eV/nm or less, the QF has a value of unity but can rise to values of 20 for fission fragments with LET values of 175 eV/nm. In certain special cases other modifying factors (like QF) may be brought in to account for non-uniform spatial distribution, dose rate effects and fractionation of the absorbed dose, but in general health physics, just the QF is used to obtain the dose equivalent.

The maximum permissible exposure rate for the general population is recommended as 0·5 rem/year and 0·08 mrem/h, where the year is the normal one of 365 days and a 24 h day. For workers with ionising radiations with respect to a 50 week year and a 40 h week the maximum permissible exposure rate is 5·0 rem/year and 2·5 mrem/h. These are the order of dose rates which can be expected around a shielded radiation facility. Any higher figures would demand remedial action in the form of restriction of access time or additional shielding, or both. Workers using ionising radiations may go up to 3 rem in a 4 month period, as long as the total for the year remains below 5 rem. Where slightly higher doses are measured, ionisation chambers may be useful. Quartz fibre pocket electroscopes can be used to measure accumulated personal dose. It is essential, and a legal requirement in the U.K., that a record be kept of this for all workers using ionising radiations.

A cheaper and more convenient method for determining personnel dose, and at the same time keeping a permanent record, is the film badge. This consists of a small piece of wrapped photographic film coated with a fast emulsion on one side and a slow emulsion on the other side. These films are numbered and worn for 2–4 weeks in a plastic holder, which contains some small pieces of tin and Dural metal. The metal acts as a filter and permits the determination of the type and energy of the radiation to which the film has been exposed. A measure of radiation exposure is obtained by comparing the optical transmission of films worn by workers with that of films which have been exposed to known doses of radiation. The films are selected from one batch and developed at the same time with the same materials. Having been processed and recorded, the films can be stored in fire-proof safes for future reference. Thus an accumulated dose record can be kept for each person working with ionising radiations.

DOSIMETRY OF INTERNAL SOURCES

The dose delivered by internal sources to solutions can be measured by a suitable chemical dosimeter. When a chemical dosimeter is used, it should be ascertained that the presence of the source, heterogeneous or homogeneous, does not interfere with the normal chemical behaviour of the dosimeter. It is also possible to calculate the dose if it is known that all the energy arising from the internal source is absorbed by the system. This is usually the case, as alpha and beta emitters account for the majority of internal sources. The total energy absorbed by the system will then be given by

$$\text{total energy absorbed} = \begin{array}{l}\text{number of disintegrations}\\ \text{during time of irradiation}\\ \times \text{ energy loss per}\\ \text{disintegration}\end{array}$$

The number of disintegrations may be obtained in the following way. It will be recalled that the law governing the decay of a radioactive substance is

$$N = N_0 e^{-\lambda t} \qquad (5.8)$$

where N is the number of atoms remaining at time t, N_0 is the number of atoms at $t = 0$ and λ is a constant characteristic of a particular radioactive species known as the decay constant. The number of disintegrations occurring in time t is thus $(N_0 - N)$ and, from equation (5.8),

$$N_0 - N = N_0 - N_0 e^{-\lambda t}$$

or

$$N_0 - N = N_0(1 - e^{-\lambda t}) \qquad (5.9)$$

The activity, A, defined as the rate of disintegration is related to N by

$$A = \lambda N \qquad (5.10)$$

or, applying this relationship at the time when $t = 0$,

$$A_0 = \lambda N_0 \qquad (5.11)$$

Substituting in equation (5.9),

$$N_0 - N = \frac{A_0}{\lambda}(1 - e^{-\lambda t}) \qquad (5.12)$$

In addition, the decay constant, λ, is related to the half-life, $t_{\frac{1}{2}}$, of the radioactive material by

$$\lambda = \frac{\ln 2}{t_{\frac{1}{2}}}$$

or

$$\lambda = \frac{0 \cdot 693}{t_{\frac{1}{2}}} \qquad (5.13)$$

Substituting in equation (5.12), the number of disintegrations in time t $(N_0 - N)$ is thus given by

$$(N_0 - N) = \frac{A_0 t_{\frac{1}{2}}}{0 \cdot 693}\left[1 - \exp\left(-\frac{0 \cdot 693 t}{t_{\frac{1}{2}}}\right)\right] \qquad (5.14)$$

NEUTRON DOSIMETRY

Neutron dosimetry presents special problems because these particles give rise to ionising radiations only indirectly by inducing radioactivity or by recoil reactions (see Chapters 2 and 3). Fast neutrons recoil mainly with hydrogen atoms, which then behave like high-speed protons in the medium under investigation. Thimble ionisation chambers working under Bragg–Gray cavity conditions can be specially adapted to measure fast protons by filling them with hydrogen and making their walls from polythene. Such a chamber can measure the dose from neutron plus gamma radiation. The latter usually accompanies neutrons. Gammas can be measured alone in a fast neutron field by leaving all hydrogeneous materials out of the gas and walls of the chamber. The dose from fast neutrons can then be obtained by difference. To obtain the dose from slow neutrons, their radioactivation of the system under investigation must be studied.

A nuclear reactor used as a radiation source gives rise to mixed fast and slow neutrons plus gamma rays. Accurate determination of the dose from high fluxes of all of these is extremely difficult and the results have sometimes been expressed, undesirably, in pile units rather than rads.

6
Water and aqueous solutions

In the previous chapters the general effects and behaviour of ionising radiations have been discussed together with the general behaviour of the free radicals which frequently result from the action of ionising radiations. These concepts may now be applied to some specific systems; water and aqueous solutions will be considered first. Although these are not the simplest systems, they are very important and certainly the most widely studied to date. In addition, many of the current ideas in radiation chemistry have developed from the study of aqueous solutions. It should be emphasised, however, that ideas on the fundamental processes in the radiolysis of aqueous solutions are still undergoing development and the topic is far from being completely understood.

THE PRODUCTS OF WATER RADIOLYSIS

In the radiolysis of water, hydrogen and hydrogen peroxide are produced. In addition to these molecular products, hydrogen atoms, H, solvated electrons, e_{aq}^-, and hydroxyl radicals, OH, have been detected by their chemical and physical properties.

When low concentrations of aliphatic solutes are added to the water, the yield of hydrogen is considerably increased. If the hydrogen atoms in the solute are replaced with deuterium, the excess hydrogen produced is HD. This could arise by virtue of the deuterium atom being abstracted from the solute by a hydrogen atom (see reaction

4.23, Chapter 4). Kinetic experiments at high ionic strengths have confirmed that the precursor of the HD gas is uncharged, and H atoms have also been identified by e.p.r. in the radiolysis of ice. Thus it has been concluded that H atoms are a product of water radiolysis.

The evidence for the solvated electron came first from observations that there appeared to be a second reducing species in radiolysed water in addition to H atoms. Sometimes this second species reacts with substrates to give the same products as H atoms, but the rates of the reactions can differ by several orders of magnitude. Sometimes the products of the reactions are different. The second species can reduce CO_2 to CO_2^-, H_3O^+ to H and N_2O to N_2, but does not react with aliphatic alcohols, ethers or amines. These last three classes of substances all undergo hydrogen abstraction reactions with H atoms. Kinetic experiments have shown that the second reducing species carries a single negative charge and this has been confirmed by conductance measurements on water subjected to pulse radiolysis. The properties of this species seem to accord with those predicted for the hydrated electron e_{aq}^-. The absorption spectrum of the second reducing species has been determined by pulse radiolysis and flash photolysis, and agrees with that predicted for e_{aq}^- from the absorption spectrum of the solvated electron in liquid ammonia. It is thus concluded that solvated electrons are also produced in the radiolysis of water.

The hydroxyl radical was originally postulated as a product in water radiolysis on the basis of oxidation reactions which were observed. Certain solutes can be oxidised in radiolysed water, provided that the products are protected from attack by the H atoms and e_{aq}^-. This is accomplished by adding other solutes which will react with these reducing species. Thus bromide can be oxidised to bromine, iodide to iodine, formate to oxalate, etc. In the pulse radiolysis of benzene, the intermediate C_6H_6OH has been identified and the oxidising species has been shown to be identical with the product arising from the photolysis of H_2O_2. It seems as though the oxidising species might be the hydroxyl radical, OH. Kinetic experiments have shown that the oxidising species is uncharged in the range pH 0–10 and dissociates into a basic form with $pK = 11\cdot7$. These facts all fit with the OH radical, which has also been identified by e.p.r. in irradiated ice.

There is also another species which is produced with high-LET radiation. This has been identified by its chemical behaviour as the perhydroxyl radical, HO_2, which has also been directly identified in frozen solutions.

In the radiolysis of water, then, the species H_2O_2, H_2, H, e_{aq}^-, OH and HO_2 are produced, and any theory of the mechanism of radiolysis must account for the occurrence of these species.

THE IONISATION PROCESS

There is no certain knowledge of the types and abundance of the ions formed in liquid water by the action of ionising radiations. Some guidance can be obtained, however, from the results of mass spectrometry of water vapour at very low pressures. Under these conditions, the most abundant ion produced is H_2O^+, the next most abundant species being OH^+, H^+ and H_3O^+. In general, there are about five times more H_2O^+ than the other ions and sometimes the quantity of H^+ can drop to about 5% of the quantity of H_2O^+. There are very small amounts of O^+, H^-, and O^- observed in the mass spectrometer. The amount of O^+ formed is negligible and the very small amounts of the negative ions indicate that the electrons removed from the water molecules in the formation of the H_2O^+ ions do not readily undergo attachment reactions with neutral species.

If the ions OH^+ and H^+ are formed in liquid water under the influence of ionising radiations, they will be formed by the reactions

$$H_2O \longrightarrow H^+ + OH + e^- \qquad (6.1)$$

$$H_2O \longrightarrow OH^+ + H + e^- \qquad (6.2)$$

An intermediate stage in reaction (6.2) may be the formation of a highly excited water molecule, H_2O^{**}. The products of the above reactions would be caged by the water molecules and might well react together according to

$$H^+ + OH \longrightarrow H_2O^+ \qquad (6.3)$$

$$OH^+ + H \longrightarrow H_2O^+ \qquad (6.4)$$

These reactions would simply produce an extra quantity of the most abundant ion, H_2O^+.

Alternatively, the H^+ ion may react with an electron to form a hydrogen atom:

$$H^+ + e^- \longrightarrow H \qquad (6.5)$$

The OH^+ ion may also react with the water molecules in either of two ways:

$$OH^+ + H_2O \longrightarrow H^+ + 2OH \qquad (6.6)$$

$$OH^+ + H_2O \longrightarrow H_2O^+ + OH \qquad (6.7)$$

The above reactions are somewhat speculative; but the important point is that if the reactions of the OH^+ and H^+ ions, as outlined above, are considered, they lead to the production of H_2O^+, OH and H. It will be proposed later that the reactions of the most abundant ion, H_2O^+, also lead eventually to the production of OH and H, among other things, so that nothing is lost if the OH^+ and H^+ ions are neglected. Although the OH^+ ion may possibly undergo other reactions, it is usually assumed that the most important ionic species formed in the radiolysis of water is the H_2O^+ ion. The ionisation process in water may thus be represented

$$H_2O \longrightarrow\!\!\!\sim\!\!\!\sim\!\!\!\sim\!\!\!\longrightarrow H_2O^+ + e^- \qquad (6.8)$$

THE EXCITATION PROCESS

Some water molecules will be raised to an excited state by the action of ionising radiations. It is usually assumed that most of these return to the ground state by a non-radiative process or dissociate according to

$$H_2O^* \longrightarrow H + OH \qquad (6.9)$$

The products of this reaction will probably be confined by a solvent cage and will recombine to produce no net chemical change. The contribution of the excited molecules to the radiolysis of water and aqueous solutions is thus generally ignored, except in a few special circumstances where the reactions of the excited molecules are invoked to explain a particular effect.

THE MECHANISM OF WATER RADIOLYSIS

In the same way that the detailed ionisation process is speculative, the reactions which immediately follow ionisation are also debatable and the situation is by no means resolved. The mechanism which accounts for most, but not all, experimental observations, and is generally accepted at the moment as the best approximation, is often called the *diffusion model*. It has gradually developed over a period of time, undergoing modification as new experimental facts came to light. Originally there were two slightly different points of view, the basic difference concerning the distance travelled by the electron ejected from the water molecule in reaction (6.8) before its energy was reduced to a thermal level.

One view put forward by Samuel and Magee suggested that most electrons ejected from the parent molecules travelled only a very small distance before they were reduced to thermal energy. It was suggested that under these circumstances the electron would still be within the electric field of the H_2O^+ ion and, as the electron would have only a small energy, it would be drawn back to the H_2O^+ ion and neutralisation would occur. It has been pointed out in Chapter 4 that such a neutralisation would lead to a highly excited state which would probably dissociate. The Samuel and Magee process can thus be represented

$$H_2O^+ + e^- \longrightarrow H_2O^{**} \longrightarrow H + OH^- \qquad (6.10)$$

The eventual products are thus OH and H. As these radicals are formed from a highly excited state, they will probably be formed with sufficient energy to escape from the solvent cage and thus avoid recombination. Samuel and Magee thus suggested that, after ionisation, H and OH radicals would be formed in close proximity to one another.

The other view was put forward by Lea and Gray, who suggested that the electron ejected in reaction (6.8) would travel a relatively great distance from the H_2O^+ ion before it was reduced to thermal energy. The important point here is that the electron would be well away from the electric field of the H_2O^+ ion when it became thermalised and that under these circumstances both the H_2O^+ ion and the electron would react independently with solvent molecules. It has been shown that the ion H_2O^+ reacts very rapidly with a water molecule to give an OH radical:

$$H_2O^+ + H_2O \longrightarrow H_3O^+ + OH^- \qquad (6.11)$$

The reaction of the electron was suggested to be

$$e^- + H_2O \longrightarrow H + OH^- \qquad (6.12)$$

The Lea–Gray idea thus leads to the same active products, H and OH, as the Samuel–Magee scheme but the difference is in the distribution of these species. On the Samuel–Magee scheme they are formed close together, while on the Lea–Gray scheme the hydrogen atoms will be formed at some distance from the hydroxyl radicals.

The Lea–Gray scheme was modified by Platzmann, who agreed that the ejected electron would travel beyond the influence of the H_2O^+ ion's electric field before it became thermalised. He pointed out, however, that reaction (6.12) is relatively slow and suggested that the electron might have time to become solvated and survive in this form to react with other active species rather than solvent molecules.

In the early days of radiation chemistry, most of the evidence for the existence of free radicals was based on chemical kinetics and the postulation of H and OH was sufficient to explain the observed effects. Since then, however, the techniques of absorption spectroscopy, conductance measurements and e.p.r., coupled with pulse radiolysis, have been applied to the field, and the existence in the radiation chemistry of aqueous solutions of H, OH and e_{aq}^- has been confirmed. Any mechanism of radiolysis must thus account for the formation of these species and the currently accepted version of the diffusion model, which dates from about 1963, does this.

According to this mechanism, the ionising particles cause ionisation and excitation of the water molecules. The excited molecules are assumed to undergo de-excitation without producing any chemical products. The ionisation process is represented by equation (6.8)

$$H_2O \longrightarrow\!\!\!\sim\!\!\!\sim\!\!\!\rightarrow H_2O^+ + e^- \qquad (6.8)$$

The half-life of this process is estimated to be about 10^{-16} s.

The ions and the electrons then react with the water according to

$$H_2O^+ + H_2O \longrightarrow H_3O^+ + OH \qquad (6.11)$$

$$e^- + aq \longrightarrow e_{aq}^- \qquad (6.13)$$

Reaction (6.11) is very rapid and the half-lives of these two reactions are estimated as 10^{-14} s and 10^{-10} s, respectively. The initial stage of the process is thus the formation of OH and e_{aq}^-. These radicals and electrons will not, of course, be distributed uniformly throughout the medium, as they are formed from the ions which are located in spurs on the track of the ionising particle. If solvated electrons are formed, the electrons ejected from the water molecule in reaction (6.8) must travel beyond the electric field of the H_2O^+ ion before they become solvated. This idea is not in conflict with the postulate that the initial positions of the radicals and solvated electrons are within the spur. It is a question of considering the distances involved. Platzmann considered that the minimum distance which would be travelled by an electron ejected from the parent molecule with an energy of 10 eV would be about 5 nm before it became thermalised. At this distance the electron is outside the H_2O^+ ion's field. If the ionising particle is a fast electron which has a low LET, the spurs formed on the track are about 500 nm apart. Thus an electron which travels 5 nm from the ionisation point may still be considered to be in the spur. If the ionising particle has a high LET, such as an α-particle, the spurs themselves are only about 1 nm apart and can be considered to overlap from the moment of formation. These points have been discussed in Chapter 3 and the situation has been illustrated in *Figure 3.12*. In this way the spatial distribution of the radicals and solvated electrons is much the same as that depicted in *Figure 3.12* for the distribution of spurs in the cases of low- and high-LET ionising particles.

The initial distribution of OH, H_3O^+ and e_{aq}^- within the spur will not be homogeneous. As explained in reference to the original Lea–Gray scheme, the OH radicals will be more concentrated at the centre of the spur. This will be the situation at about 10^{-10} s after the ionising particle has passed. At this time, in the spurs or, for high-LET radiation, in the columnar track, the concentration of OH, H_3O^+ and e_{aq}^- is high and reactions between these species are favoured. The initial reactions (6.8), (6.11) and (6.13) are followed by

$$e_{aq}^- + H_3O^+ \longrightarrow H + H_2O \qquad (6.14)$$
$$k = 2{\cdot}4 \times 10^{10} \text{ mol}^{-1} \text{ dm}^3 \text{ s}^{-1}$$

$$e_{aq}^- + OH \longrightarrow OH^- \qquad (6.15)$$
$$k = 3{\cdot}0 \times 10^{10} \text{ mol}^{-1} \text{ dm}^3 \text{ s}^{-1}$$

$$e_{aq}^- + e_{\,q} \longrightarrow H_2 + 2OH^- \qquad (6.16)$$
$$k = 4{\cdot}5 \times 10^9 \text{ mol}^{-1} \text{ dm}^3 \text{ s}^{-1}$$

$$OH + OH \longrightarrow H_2O_2 \qquad (6.17)$$
$$k = 5{\cdot}0 \times 10^9 \text{ mol}^{-1} \text{ dm}^3 \text{ s}^{-1}$$

where the rate constants refer to the appearance of the products.

The above reactions can then be followed by reactions involving the hydrogen atom, the product of reaction (6.14), with hydrated electrons and hydroxyl radicals

$$e_{aq}^- + H \longrightarrow H_2 + OH^- \qquad (6.18)$$
$$k = 2{\cdot}5 \times 10^{10} \text{ mol}^{-1} \text{ dm}^3 \text{ s}^{-1}$$

$$OH + H \longrightarrow H_2O \qquad (6.19)$$
$$k = 7 \times 10^9 \text{ mol}^{-1} \text{ dm}^3 \text{ s}^{-1}$$

As time goes on, the spurs and tracks expand by diffusion. The radical concentration thus decreases and some of the original radicals are able to diffuse into the bulk of the solution before reacting. The stable products, H_2O_2 and H_2, will also diffuse into the bulk of the solution. The mathematical treatment of this diffusion model predicts the formation of H_2, H_2O_2 and H in the spurs, and the model is most successful in accounting for the effects of LET on the relative quantities of the stable products and the radicals which escape from the spurs or tracks into the bulk of the solution.

It will be appreciated on the basis of the above model that as the H_2O_2 and H_2 are formed from radical–radical reactions in the spurs, the greatest yields of these products would be expected from situations giving rise to the greatest concentrations of radicals. With low-LET radiation, the spurs are more or less isolated and the radicals can diffuse away from one another relatively freely, so that the radical concentration decreases quite rapidly. In this situation it is expected that the amounts of H_2O_2 and H_2 formed in the spurs will be relatively small and the amounts of radicals escaping into the bulk of the solution will be relatively large. With high-LET radiation, on the other hand, the spurs overlap from the moment of formation and the radicals are densely packed in a cylindrical zone. Under these conditions there will be many radical–radical reactions, forming H_2 and H_2O_2 and only a few radicals will escape into the bulk of the solution. With high-LET radiation, therefore, it would be expected that larger quantities of H_2 and H_2O_2 will be formed in the track than in the case of low-LET radiation. Accordingly, it would also be expected that the number of radicals escaping into the bulk of the solution will be much lower with high-LET radiation than with low-LET radiation.

In the tracks of high-LET particles where the radical concentration is particularly high, the H_2 and H_2O_2 formed by reactions (6.16), (6.17) and (6.18) may react further with radicals before escaping into the bulk of the solution, where the radical concentration will be much lower:

$$e_{aq}^- + H_2O_2 \longrightarrow OH + OH^- \qquad (6.20)$$
$$k = 1\cdot2 \times 10^{10} \ \text{mol}^{-1} \ \text{dm}^3 \ \text{s}^{-1}$$

$$H + H_2O_2 \longrightarrow OH + H_2O \qquad (6.21)$$
$$k = 9 \times 10^7 \ \text{mol}^{-1} \ \text{dm}^3 \ \text{s}^{-1}$$

$$OH + H_2O_2 \longrightarrow HO_2 + H_2O \qquad (6.22)$$
$$k = 2\cdot3 \times 10^7 \ \text{mol}^{-1} \ \text{dm}^3 \ \text{s}^{-1}$$

$$OH + H_2 \longrightarrow H + H_2O \qquad (6.23)$$
$$k = 6 \times 10^7 \ \text{mol}^{-1} \ \text{dm}^3 \ \text{s}^{-1}$$

It will be noticed that these last three reactions in particular are much slower than the others which have been discussed, and will thus only occur to any extent in a spur or track where the concentrations of radicals and H_2O_2 or H_2 are large. Of these reactions, only (6.22) gives rise to a new product, the perhydroxyl radical, HO_2, and this is usually formed only in very small quantities in the tracks of high-LET radiation.

H

RADICAL AND MOLECULAR YIELDS

It will be appreciated from the foregoing that the detailed kinetics of radiation chemistry are quite complex as the situation is far removed from that obtaining in a homogeneous system. Interest centres on the radiation chemistry of aqueous solutions and so the effect of the presence of solutes must be considered. Progress can be made by a simplification resulting from the use of dilute solutions, but it is essential to appreciate the limitations imposed by such a simplification. The position may be understood from a considera-tion of the competition between radical–radical and radical–solute reactions.

The hydroxyl radical is, of course, an oxidising species, and the hydrogen atom and the solvated electron can act as reducing species. If an oxidisable substrate is introduced into the water, it will tend to react with OH. Such a reaction will compete with other reactions in which OH is involved. For example, if some substrate S were oxidised to S^+ in an electron-transfer reaction with OH such as

$$S + OH \longrightarrow S^+ + OH^- \qquad (6.24)$$

this reaction would compete with reactions (6.15), (6.17) and (6.19). A substance which removes radicals in this way is frequently called a *scavenger*. In the presence of the scavenger, S, the amount of H_2O_2 formed would be decreased and the amount of OH available for reaction with S could be computed from the yield of S^+. If the concentration of S were increased, reaction (6.24) would occur more rapidly and divert more OH from the formation of H_2O_2. The yield of H_2O_2 would decrease further and as the yield of S^+ would increase it would appear that more OH was available for reaction with S. The same considerations apply to the situation where a reducible substrate would compete for H and e_{aq}^-.

It is important to realise when such competition as has been described above is important and when it is negligible. A hypo-thetical example should illustrate the point. Suppose that radicals $R\cdot$ can be generated in a system and that these radicals can either dimerise to give a stable product, R_2, or oxidise a substrate in an electron transfer reaction:

$$R\cdot + R\cdot \longrightarrow R_2 \qquad (6.25)$$

$$R\cdot + S \longrightarrow S^+ + R^- \qquad (6.26)$$

If these were the only reactions in which the radicals could take part, then the fraction of them, f, taking part in reaction (6.26) is given by

$$f = \frac{v_{26}}{v_{26} + v_{25}}$$

where v_{25} and v_{26} are the velocities of reactions (6.25) and (6.26), respectively. Expressing this fraction in terms of rate constants and the concentrations of reactants,

$$f = \frac{k_{26}[S]}{k_{26}[S] + k_{25}[R\cdot]}$$

It would be reasonable to suppose that $k_{25} = 10^2 \times k_{26}$, so that the fraction of radicals scavenged by the solute becomes

$$f = \frac{[S]}{[S] + 10^2[R\cdot]} \qquad (6.27)$$

At the time of formation of a spur in the radiolysis of aqueous solutions, the concentration of radicals within the spur is of the order of 10^{-1} mol dm^{-3}. Substituting $[R\cdot] = 10^{-1}$ mol dm^{-3} and putting $[S] = 10^{-3}$ mol dm^{-3} in equation (6.17) gives

$$f = \frac{10^{-3}}{10^{-3} + 10} \approx 10^{-4}$$

Under these conditions, the fraction of radicals being scavenged within the spur is negligible. Of course, as the spurs expand, the radical concentration decreases rapidly and in the bulk of a solution subjected to radiolysis the radical concentration is probably of the order of 10^{-10} mol dm^{-3}. Substituting $[R\cdot] = 10^{-10}$ mol dm^{-3} and putting $[S] = 10^{-6}$ mol dm^{-3} in equation (6.27) gives

$$f = \frac{10^{-6}}{10^{-6} + 10^{-10}} \approx 1$$

and under these conditions all the radicals would be scavenged by the solute.

Although this is an oversimplified consideration of the situation obtaining in radiolysis, it illustrates that there are two extreme situations. In the early life of the spur, radical–radical reactions will occur unhindered by the presence of any active solute unless it is present in concentrations greater than about 10^{-3} mol dm^{-3}. As the spur expands, the products of the radical reactions, H_2 and H_2O_2, together with any unreacted radicals diffuse into the bulk of the solution. Here radical–solute reactions occur, reactions between radicals being completely suppressed if the active solute is present in concentrations of 10^{-6} mol dm^{-3} or greater.

The H_2 and H_2O_2 which diffuse from the spurs are called the *molecular products* and their radiolytic yields are known as the *molecular yields*, denoted by G_{H_2} and $G_{H_2O_2}$, respectively. The yields of H, e_{aq}^- and OH which escape from the spurs are called the *radical yields*, and are denoted by G_H, G_{e^-} and G_{OH}, respectively. It should be noted that in these cases the product is appended as a subscript to the symbol G, and these G values represent the earliest detectable yields of the various products as opposed to a symbol such as $G(H_2O_2)$, which would represent the final yield of H_2O_2 at the conclusion of irradiation. The application of the concepts of molecular and radical yields in dilute solutions will be discussed later in the chapter.

Provided that any scavengers present in an aqueous system are of a sufficiently low concentration not to interfere with the radical–radical reactions in the spurs, the effect of radiolysis of such a system is to produce in the bulk of the solution the molecular products, H_2 and H_2O_2, and the radical products, H, OH, e_{aq}^- and HO_2. This effect is usually represented

$$H_2O \longrightarrow\!\!\!\!\sim\!\!\sim\!\!\sim\!\!\sim\!\!\longrightarrow H_2O_2, H_2, H, OH, e_{aq}, HO_2 \qquad (6.28)$$

although it should be remembered that HO_2 radicals are produced only as a primary product from high-LET radiation. Moreover, although it has been indicated that the molecular products arise from the dimerisation of radicals in the spurs, there have been suggestions that some of the molecular products may arise from excited water molecules by

$$H_2O^* + H_2O^* \longrightarrow H_2O_2 + H_2 \qquad (6.29)$$

This reaction, however, is not currently considered to be very plausible.

Frequently, the products of reaction (6.28) are referred to as the *primary products* of water radiolysis. In fact, the very first products are, of course, ions, electrons and excited species:

$$H_2O \longrightarrow\!\!\!\!\sim\!\!\sim\!\!\sim\!\!\sim\!\!\longrightarrow H_2O^+, e^-, H_2O^* \qquad (6.30)$$

this process occupying about 10^{-15} s. The next stage comprises the ion–molecule reactions, the hydration of the electrons and the dissociation or return to the ground state of the excited states. The total time occupied up to this stage is about 10^{-10} s and the over-all process may be represented

$$H_2O \longrightarrow\!\!\!\!\sim\!\!\sim\!\!\sim\!\!\sim\!\!\longrightarrow OH, e_{aq}^- \qquad (6.31)$$

After this, the radicals and hydrated electrons react together to give H_2, H, H_2O_2 and HO_2, the over-all process being represented by reaction (6.28), the time necessary to reach this point being about 10^{-8} s. For the radicals produced to react with a scavenger at a concentration of about 1 mol dm^{-3}, the time required is about 10^{-5} s, so that it is clear that in dilute solutions the solute cannot compete with any process up to the stage represented by reaction (6.28). Thus as far as the solute is concerned, the first products which it 'sees' are the molecular and radical products diffusing from the spurs. It is for this reason that these products may be called the primary products.

RADIOLYSIS OF PURE WATER

The effects of ionising radiation on pure water depend to some extent upon the experimental conditions and the nature of the radiation. As a result of radiolysis, the molecular and radical products formed in the spurs or tracks diffuse into the bulk of the solution and, in the absence of any solute, the radicals can react with the molecular products. The possible reactions are (6.20) – (6.23):

$$e_{aq}^- + H_2O_2 \longrightarrow OH + OH^- \qquad (6.20)$$
$$k = 1 \cdot 2 \times 10^{10} \text{ mol}^{-1} \text{ dm}^3 \text{ s}^{-1}$$

$$H + H_2O_2 \longrightarrow OH + H_2O \qquad (6.21)$$
$$k = 9 \times 10^7 \text{ mol}^{-1} \text{ dm}^3 \text{ s}^{-1}$$

$$OH + H_2O_2 \longrightarrow HO_2 + H_2O \qquad (6.22)$$
$$k = 2 \cdot 3 \times 10^7 \text{ mol}^{-1} \text{ dm}^3 \text{ s}^{-1}$$

$$OH + H_2 \longrightarrow H + H_2O \qquad (6.23)$$
$$k = 6 \times 10^7 \text{ mol}^{-1} \text{ dm}^3 \text{ s}^{-1}$$

It will be noticed that H_2O_2 may be destroyed by the first three of these reactions. Reaction (6.22), however, is slower than the first two reactions; and unless there is a relatively large amount of H_2O_2 present, reaction (6.22) may be neglected. This leaves the OH radicals free to attack the H_2. With low-LET radiation, where the molecular yield of H_2O_2 is low, the effective reactions are:

$$e_{aq}^- + H_2O_2 \longrightarrow OH + OH^- \qquad (6.20)$$

$$H + H_2O_2 \longrightarrow OH + H_2O \qquad (6.21)$$

$$OH + H_2 \longrightarrow H + H_2O \qquad (6.23)$$

In all these reactions a radical and a molecular product are replaced by a radical and water (or, in the case of reaction 6.20, a hydroxide ion which will combine with a hydrogen ion to form water). The OH and H radicals produced can combine to form H_2O_2, H_2 and H_2O but the over-all process is the destruction of H_2O_2 and H_2. Thus molecular products are formed in the spurs and are destroyed in the bulk of the solution. The net result is to produce a stationary state, where the water contains small stationary concentrations of the molecular products. The lowest-LET radiations, such as γ-rays, produce very little change in pure water but with slightly higher-LET radiation the stationary concentrations of the molecular products are measurable. With high-LET radiation, the yields of molecular products diffusing from the tracks are much higher, and the yields of radicals are correspondingly smaller. In the bulk of the solution there are not enough radicals to decompose the molecular products at any appreciable rate and the latter continue to accumulate in solution, so that the water is continuously decomposed.

In an open system where the hydrogen gas can escape, the other molecular product, H_2O_2, builds up in solution to a greater extent than in a closed system. Eventually, reaction (6.22) becomes important and the HO_2 radicals formed by this reaction can react further in two ways:

$$HO_2 + HO_2 \longrightarrow H_2O_2 + O_2 \qquad (6.32)$$

$$HO_2 + OH \longrightarrow H_2O + O_2 \qquad (6.33)$$

The H_2O_2 is thus being destroyed and once again reaches a stationary concentration when its rate of production is equal to its rate of destruction. In open systems, however, the stationary concentration is rather higher than in closed systems. The molecular product, hydrogen, is escaping from the solution and oxygen is formed by reactions (6.32) and (6.33). The radiation is thus effectively decomposing water into hydrogen and oxygen.

THE EFFECT OF OXYGEN

Oxygen dissolved in water scavenges H atoms very effectively to form HO_2 radicals:

$$H + O_2 \longrightarrow HO_2 \qquad (6.34)$$
$$k = 2 \times 10^{10} \text{ mol}^{-1} \text{ dm}^3 \text{ s}^{-1}$$

It can also scavenge hydrated electrons:

$$e_{aq}^- + O_2 \longrightarrow O_2^- \qquad (6.35)$$
$$k = 1 \cdot 9 \times 10^{10} \text{ mol}^{-1} \text{ dm}^3 \text{ s}^{-1}$$

The O_2^- ion is the alkaline analogue of the HO_2 radical, which can behave as an acid and undergo dissociation. In fact, the HO_2 radicals formed in reaction (6.34) will probably be dissociated in pure water (pH 7):

$$HO_2 \longrightarrow H^+ + O_2^- \qquad (6.36)$$

The presence of oxygen in water thus leads to the conversion of H and e_{aq}^- to O_2^-. In this way, reactions (6.20) and (6.21) are suppressed and the final yield of H_2O_2 is enhanced by combination of two O_2^- ions in a reaction which is the alkaline analogue of reaction (6.32):

$$O + O_2^- \xrightarrow{\text{(2H}_2\text{O)}} H_2O_2 + O_2 + 2OH^- \qquad (6.37)$$
$$k = 1 \cdot 7 \times 10^7 \text{ mol}^{-1} \text{ dm}^3 \text{ s}^{-1}$$

THE EFFECT OF pH ON THE PRIMARY SPECIES

It has been mentioned above that the HO_2 radical exists in an acidic form and an alkaline form. Strictly, reaction (6.36) should be written as an equilibrium:

$$HO_2 \rightleftharpoons H^+ + O_2^- \qquad pK = 4 \cdot 4 \qquad (6.38)$$

Thus in acid solution the perhydroxyl radical will exist largely as HO_2, but in neutral and alkaline solution it will be present mostly in the form O_2^-. This can have an important bearing on the reactions occurring in aqueous systems. For example, HO_2 will oxidise ferrous ions but O_2^- acts as a reducing agent for ferric ions:

$$Fe^{2+} + HO_2 \longrightarrow Fe^{3+} + HO_2^- \qquad (6.39)$$
$$Fe^{3+} + O_2^- \longrightarrow Fe^{2+} + O_2 \qquad (6.40)$$

Even if the different forms of a particular species lead to the same reaction product, the speeds of the reactions may differ and this may have an important result in a mechanism involving competitive reactions.

The other primary products are also affected by pH. Hydrogen peroxide may also be regarded as an acid, its dissociation being represented

$$H_2O_2 \rightleftharpoons H^+ + HO_- \qquad pK = 11 \cdot 75 \qquad (6.41)$$

In acid and neutral solutions hydrogen peroxide exists as H_2O_2 but in alkaline solution it is present as HO_2^-. Thus, whereas in neutral solutions ferrocyanide is slowly oxidised by peroxide, in alkaline solutions ferricyanide is rapidly reduced:

$$Fe(CN)_6^{4-} + H_2O_2 \longrightarrow Fe(CN)_6^{3-} + OH + OH^- \qquad (6.42)$$
$$2Fe(CN)_6^{3-} + HO_2^- \longrightarrow 2Fe(CN)_6^{4-} + H^+ + O_2 \qquad (6.43)$$

The hydroxyl radical also dissociates:

$$OH \rightleftharpoons H^+ + O^- \qquad pK = 11 \cdot 7 \qquad (6.44)$$

so that in alkaline solutions it exists in the form O^-.

It is interesting here to note how the change in form affects the speeds of reactions by considering reaction (6.23) with its alkaline analogue:

$$OH + H_2 \longrightarrow H + H_2O \qquad (6.23)$$
$$k = 6 \times 10^7 \text{ mol}^{-1} \text{ dm}^3 \text{ s}^{-1}$$

$$O^- + H_2 \longrightarrow H + OH^- \qquad (6.45)$$
$$k = 1{\cdot}6 \times 10^8 \text{ mol}^{-1} \text{ dm}^3 \text{ s}^{-1}$$

The contrast is even more striking in the reaction between hydroxy radicals and hydrogen peroxide, where both reactants have a different form in alkaline solution:

$$OH + H_2O_2 \longrightarrow HO_2 + H_2O \qquad (6.22)$$
$$k = 2{\cdot}3 \times 10^7 \text{ mol}^{-1} \text{ dm}^3 \text{ s}^{-1}$$

$$O^- + HO_2^- \longrightarrow O_2^- + OH^- \qquad (6.46)$$
$$k = 9 \times 10^8 \text{ mol}^{-1} \text{ dm}^3 \text{ s}^{-1}$$

The remaining two primary products, H and e_{aq}^-, are also affected by pH, in so far as they react directly with hydrogen ions and hydroxide ions. The reaction of e_{aq}^- with a hydrogen ion, which has already been postulated as occurring in spurs, can occur in the bulk of the solution at low pH:

$$e_{aq}^- + H_3O^+ \longrightarrow H + H_2O \qquad (6.14)$$

In acid solutions, most, if not all the hydrated electrons, will thus be converted to hydrogen atoms. In alkaline solutions, on the other hand, the hydrogen atoms emerging from the spurs will react with hydroxide ions to become hydrated electrons:

$$H + OH^- \longrightarrow e_{aq}^- \qquad (6.47)$$

In alkaline solutions, then, the primary hydrogen atoms are rapidly converted to hydrated electrons.

Once again, these effects can affect the speeds of reactions and the contrast may be seen from a comparison of the speed of reaction (6.21) which will occur in acid solution with that of its alkaline analogue:

$$H + H_2O_2 \longrightarrow OH + H_2O \qquad (6.21)$$
$$k = 9 \times 10^7 \text{ mol}^{-1} \text{ dm}^3 \text{ s}^{-1}$$

$$e_{aq}^- + HO_2^- \longrightarrow O^- + OH^- \qquad (6.48)$$
$$k = 3{\cdot}5 \times 10^9 \text{ mol}^{-1} \text{ dm}^3 \text{ s}^{-1}$$

The interconversion of e_{aq}^- and H in acid or alkaline solutions can also give rise to different reaction products under different conditions. This is illustrated by the classic example of monochloracetic acid solutions. In acid solutions one of the major products of radiolysis of monochloracetic acid is hydrogen gas, which is attributed to the abstraction of hydrogen from the molecule by hydrogen atoms:

$$H + CH_2ClCOOH \longrightarrow H_2 + CHClCOOH \qquad (6.49)$$

In neutral solutions, on the other hand, the hydrogen yield is decreased, the deficit being made up by the production of chloride ions owing to the reaction of hydrated electrons with monochloracetic acid:

$$e_{aq}^- + CH_2ClCOOH \longrightarrow Cl^- + CH_2COOH \qquad (6.50)$$

These radiolytic observations are supported by experiments in which hydrogen atoms have been generated in an electric discharge and passed into solutions of monochloracetic acid. Between pH values of 4 and 10, the main reaction is a hydrogen abstraction reaction but above pH 11 it appears that the hydrogen atoms are converted to hydrated electrons, as the chloride ion is then the major product.

From the point of view of radiolysis, both H and e_{aq}^- emerge from the spurs as primary products. In neutral solutions both species will react with solutes. In acid solutions the concentration of hydrogen ions will be much greater than that of a solute present at a concentration of 10^{-3} mol dm^{-3} and all the e_{aq}^- will be converted to H by reaction (6.14) before they can react with the solute. In alkaline solutions, the concentration of hydroxide ions will be large enough to convert all the H to e_{aq}^- by reaction (6.47). These situations represent extreme cases. The rates of reactions (6.14) and (6.47) will, of course, depend upon pH. In mildly alkaline solutions reaction (6.47) will occur to some extent, but not exclusively, and the same may be said of reaction (6.14) in mildly acid solutions.

The fact that whereas hydrated electrons do not abstract hydrogen from solutes, hydrogen atoms do, is probably the most important difference in the chemical behaviour of H and e_{aq}^-. This is relevant to the radiolysis of deoxygenated aqueous solutions of all hydrogen-containing organic compounds. In these systems, at low pH, hydrogen abstraction reactions will predominate but become decreasingly important as the pH rises.

There is one other pH-dependent equilibrium which has been proposed. This is the association of hydrogen atoms with hydrogen ions:

$$H + H^+ \rightleftharpoons H_2^+ \qquad pK \approx 2 \cdot 5 \qquad (6.51)$$

This reaction has been proposed to explain the observation that in certain cases hydrogen atoms in acid solutions appear to behave as oxidising agents. There are, however, other explanations which could fit the observed effects. The forward reaction is considered to be relatively slow.

LIMITATIONS OF THE CONCEPTS OF RADICAL AND MOLECULAR YIELDS

It will be realised that the molecular yields cannot be determined in pure water, as the molecular products are attacked by the radical products. This has been shown in the section on pure water. In fact, the radical and molecular yields in the radiation chemistry of aqueous solutions are determined from the effects of irradiation upon active solutes present in the system. These methods will be considered in greater detail in the next section, but it will be useful at this stage to establish the significance and limitations of these yields.

It has been pointed out already that there are two extreme situations, one obtaining in a spur at the moment of its formation and the other obtaining in the bulk of the solution. There is, however, no sharp boundary between these two situations; they gradually merge into each other. In the absence of any scavengers in solution—that is, with absolutely pure water—the radicals which diffuse from the spurs into the bulk of the solution eventually react with one another or with the molecular products which have also diffused from the spurs. If a scavenger is added to the water, it will react with some of the radicals, and it has already been mentioned that only very low concentrations of solutes are required to scavenge all the radicals in the bulk of the solution. As the scavenger concentration is increased, it will first scavenge all the radicals in the bulk of the solution and then start to compete with radical–radical reactions which are occurring within the spur, but which occur late in the life of the spur when it has expanded to some considerable extent, and thus when the radical concentrations in this region are somewhat lowered. Eventually the scavenger concentration will reach a level when it begins to interfere with radical–radical reactions which are occurring at an earlier stage in the life of the spur.

It will be appreciated that, as the solute starts to scavenge radicals within the spur, there will be less radical–radical reactions forming the molecular products. If the solute scavenges OH, then $G_{H_2O_2}$ will appear to decrease and, as judged by the products of the scavenging reaction, G_{OH} will appear to increase. Conversely, if the solute scavenges H or e_{aq}^-, then G_{H_2} will appear to decrease and G_H or G_{e^-} will appear to increase.

These effects are shown in *Figure 6.1*, in which the yields of the molecular products relative to their yields at infinite dilution are plotted against solute concentration for several cases of γ-radiolysis. The results of the various experiments with different solutes have been multiplied by an arbitrary factor so that all the results lie on a single curve. It is of interest to note that $G_{H_2O_2}$ (relative) and G_{H_2} (relative) both vary in the same manner with solute concentration, which is some evidence for believing that both primary molecular products arise in a smiliar way from radical–radical reactions.

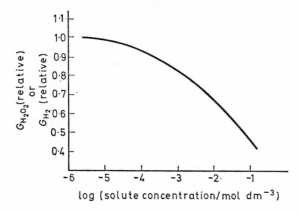

Figure 6.1.   Variation of molecular yields with solute concentration

It can be understood from *Figure 6.1* that for dilute solutions of concentrations between 10^{-6} mol dm^{-3} and 10^{-3} mol dm^{-3} and with radiations of low-LET such as γ-rays, X-rays and fast electrons, the molecular and radical yields vary only very slightly with scavenger concentration. In many cases the yields may be considered independent of scavenger concentration under these conditions or alternatively, as the variation is slight, they are readily extrapolated to infinite dilution. With low LET radiation the spurs are essentially isolated and the radicals can diffuse outwards in three dimensions. In this case the radical concentration will drop very rapidly with distance from the centre of the spur and it is not too difficult to distinguish between the region within the spur and the region outside the spur where the radicals can be scavenged. In this situation there is some justification for the concept of a radical yield.

With high-LET radiation, on the other hand, the radicals are formed in a cylindrical track and can only diffuse outwards in two dimensions. It can be appreciated that here the concentration of radicals drops much less sharply with distance from the centre of the track. In this situation it is rather more difficult to answer the question, 'When is a spur not a spur?' If the concentration of radicals drops only gradually from the centre of the track, then increasing concentrations of scavengers will be gradually more effective.

Although these considerations are only qualitative, there is no doubt in practice that radical yields determined for high-LET radiation, such as α-particles, are much more sensitive to scavenger concentration than those determined for low-LET radiation. Thus values of radical and molecular yields obtained with high-LET radiation should be regarded with considerable reserve and are generally less useful than low-LET radical and molecular yields. Even the molecular and radical yields obtained for low-LET radiation can be regarded as only approximately constant for scavenger concentrations between 10^{-5} mol dm^{-3} and 10^{-3} mol dm^{-3}.

Most work in the field of radiation chemistry is done with low-LET radiations and the concentration ranges studied are frequently kept within the above limits, so that the concepts of radical and molecular yields are useful ones. It must be remembered, however, in considering the results of work carried out under other conditions, that radical and molecular yields in aqueous systems are not constant.

RADIOLYSIS OF AQUEOUS SOLUTIONS

Before we consider methods of determination of radical and molecular yields it is appropriate to make some observations on the general experimental aspects of radiolysis.

EXPERIMENTAL CONDITIONS

Something has already been said in Chapter 2 about the vessels used for the irradiation of various systems. In addition, it has been pointed out in the present chapter that concentrations of 10^{-6} mol dm^{-3} are enough to scavenge all the radicals in the bulk of the solution. It will thus be appreciated that great care must be taken to remove impurities from any water which is to be used in radiolysis studies. Water is commonly triply distilled, ordinary distilled water being redistilled from alkaline permanganate solution and then from acid dichromate solution. The alkaline and acid solutions may be refluxed for long periods and the final product may be pre-irradiated before being used as the solvent for the system under study, in order to remove the final traces of any organic impurity.

YIELD-DOSE GRAPHS

Normally an aqueous solution will be allowed to absorb a given dose of radiation, usually by being irradiated for a given time at a constant dose rate, and the products of reaction will be determined by the usual chemical methods. The experiment will be repeated for different doses and a graph of the amount of any particular product against the dose absorbed may be constructed. The G-value for this product will be given by the slope of the graph. If the graph is linear through the origin, this determination is straightforward, the G-value being constant over the period investigated. As the irradiation proceeds, the reactant in the solution is being consumed and its concentration is therefore decreasing. A linear yield–dose curve thus indicates that the G-value is independent of concentration over the range studied. Moreover, it also indicates that the products of reaction, which must build up in the solution during irradiation, are not affecting the rate of production of the particular product in which interest lies. This may be either because the products of reaction are stable, or because the dose absorbed is kept low so that the concentration of product is low and any reactions in which it takes part are very slow and negligible.

If the yield–dose graph is a curve, the implication is that the G-value is dependent upon the concentration of the reactant or that the product is being consumed in some side or back reactions as its concentration builds up. In this case, the G-value for the known initial conditions is given by the tangent to the yield–dose curve at the origin.

MATERIAL BALANCE EQUATIONS

Between the yields of the primary products of radiolysis and the net yield of water decomposed there must exist a material balance. This can be deduced by setting the number of hydrogen atoms in each product formed per 100 eV of energy absorbed equal to the number of hydrogen atoms in the net water decomposed for the same amount of absorbed energy. The net water decomposed per 100 eV of energy absorbed to form primary products may be denoted G_{-H_2O}. The amounts of primary products formed per 100 eV of energy absorbed are given by G_{H_2}, $G_{H_2O_2}$, G_H, G_{e^-}, G_{OH} and G_{HO_2}. In one molecule of water there are two hydrogen atoms; in one molecule of H_2 there are two; in one molecule of H_2O_2 there are two; in one molecule of OH there is one; and in one molecule of HO_2 there is one. The hydrated electron is stoichiometrically equivalent to a hydrogen atom and, summarising the above statement in an equation,

$$2G_{-H_2O} = 2G_{H_2} + 2G_{H_2O_2} + G_H + G_{e^-} + G_{OH} + G_{HO_2} \quad (6.52)$$

If the same operation is carried out and the number of oxygen atoms in the net water decomposed is equated with the number occurring in the products, the corresponding relation is

$$G_{-H_2O} = 2G_{H_2O_2} + G_{OH} + 2G_{HO_2} \quad (6.53)$$

Eliminating G_{-H_2O} from equations (6.52) and (6.53),

$$2G_{H_2O_2} + G_{OH} + 3G_{HO_2} = 2G_{H_2} + G_H + G_{e^-} \quad (6.54)$$

By comparing equations (6.53) and (6.54) it can be seen that the left-hand side of the latter exceeds G_{-H_2O} by an amount equal to G_{HO_2}. Thus, if G_{HO_2} is subtracted from both sides of equation (6.54), each side will be equal to $G_{H_2O_2}$, i.e.,

$$G_{-H_2O} = 2G_{H_2O_2} + G_{OH} + 2G_{HO_2} = 2G_{H_2} + G_H + G_{e^-} - G_{HO_2} \quad (6.55)$$

For low-LET radiation where G_{HO_2} is virtually zero, equation (6.55) takes the form

$$G_{-H_2O} = 2G_{H_2O_2} + G_{OH} = 2G_{H_2} + G_H + G_{e^-} \quad (6.56)$$

These material balance equations can be useful in the determination of the yields of primary products. Obviously, it is only necessary to determine all but one of the G-values experimentally, as the other may be calculated from the material balance equation.

DETERMINATION OF RADICAL AND MOLECULAR YIELDS

Radical and molecular yields may be determined from the amount of chemical reaction produced in aqueous solutions by a known dose of radiation, provided that the detailed reaction mechanisms are known. A few examples of some of the systems studied by this 'classical' approach will be considered in this section together with some more recent methods.

ACID SOLUTIONS

For the determination of radical and molecular yields in acid solutions, cationic systems have been frequently employed.

Ferrous sulphate

The oxidation of solutions of ferrous sulphate in dilute sulphuric acid has been widely studied and provides the basis of the Fricke dosimeter mentioned in Chapter 5. The concentration of acid in the solutions is usually between 0·01 and 0·4 mol dm^{-3} H_2SO_4. Under these conditions, the primary hydrated electrons will be converted to hydrogen atoms by reaction (6.14):

$$e^-_{aq} + H_3O^+ \longrightarrow H + H_2O \qquad (6.14)$$

It is thus not possible to determine G_H and G_{e^-} separately, but only the composite term $G_H + G_{e^-}$. The hydroxyl radicals will oxidise ferrous ions according to

$$OH + Fe^{2+} \longrightarrow Fe^{3+} + OH^- \qquad (6.57)$$

In fact, it has been shown by the occurrence of H_2SO_5 and $H_2S_2O_8$ in irradiated solutions of sulphuric acid that OH radicals will also react with the acid:

$$OH + HSO^- \longrightarrow HSO_4 + OH^- \qquad (6.58)$$

The velocity constants for reactions (6.57) and (6.58) are 3×10^8 mol^{-1} dm^3 s^{-1} and 10^5 mol^{-1} dm^3 s^{-1}, respectively, so that if the concentration of ferrous ions is 10^{-2} mol dm^{-3} or greater, reaction (6.58) cannot compete to any significant extent with reaction (6.57). If reaction (6.58) does intervene to any appreciable extent, it will not affect the over-all scheme, because the HSO_4 radicals formed will also oxidise a ferrous ion:

$$HSO_4 + Fe^{2+} \longrightarrow Fe^{3+} + HSO^- \qquad (6.59)$$

Thus each OH causes the oxidation of one Fe^{2+}.

The primary H_2O_2 will also oxidise ferrous ions according to

$$H_2O_2 + Fe^{2+} \longrightarrow Fe^{3+} + OH + OH^- \qquad (6.60)$$

The OH radical formed in the last reaction will, of course, oxidise a further ferrous ion. In the presence of oxygen, the hydrogen atoms which emerge from the spurs together with those formed from hydrated electrons will combine with oxygen to form HO_2 radicals according to reaction (6.34):

$$H + O_2 \longrightarrow HO_2 \tag{6.34}$$

The HO_2 radicals so formed will then oxidise ferrous ions:

$$HO_2 + Fe^{2+} \longrightarrow Fe^{3+} + HO_2^- \tag{6.39}$$

This reaction is followed by

$$HO_2^- + H_3O^+ \longrightarrow H_2O_2 + H_2O \tag{6.61}$$

The H_2O_2 formed will oxidise two further ferrous ions by reactions (6.60) and (6.57).

From the above reaction scheme it can be seen that each molecule of primary H_2O_2 will provide two molecules of Fe^{3+} and each primary OH will provide one molecule of Fe^{3+}. The primary H and e_{aq}^- will each oxidise three molecules of Fe^{2+}. The total yield of Fe^{3+} in the presence of oxygen, $G(Fe^{3+})_{O_2}$ may thus be expressed as

$$G(Fe^{3+})_{O_2} = 2G_{H_2O_2} + G_{OH} + 3(G_H + G_{e^-}) \tag{6.62}$$

As has been indicated in Chapter 5, the Fe^{3+} in the solution at the conclusion of irradiation can be determined spectrophotometric-ally and $G(Fe^{3+})_{O_2}$ can be calculated from a yield–dose curve. For equation (6.62) to be valid, it is, of course, necessary for there to be an adequate supply of oxygen present. As oxygen is consumed during the course of the radiolysis, the dose of absorbed energy in any particular experiment must be kept below that which will consume all the oxygen dissolved in solution. For low-LET radiation, this dose is about 5×10^4 rad. If this dose is exceeded, the slope of the yield–dose curve will decrease. Another effect which will cause the yield–dose curve to depart from linearity arises on prolonged radio-lysis. The concentration of Fe^{3+} builds up and these ions begin to compete with oxygen for H atoms:

$$Fe^{3+} + H \longrightarrow Fe^{2+} + H^+ \tag{6.63}$$

In addition, the Fe^{3+} ions will also compete for HO_2:

$$Fe^{3+} + HO_2 \longrightarrow Fe^{2+} + H^+ + O_2 \tag{6.64}$$

Both these effects will decrease the yield of Fe^{3+} and cause the slope of the yield–dose curve to fall off.

If the ferrous sulphate solution is degassed, so that no oxygen is present, reaction (6.34) cannot occur and the H atoms must react in some other way. Experiments in which H atoms have been generated in electric discharges and passed into acid ferrous sulphate solutions have established that H atoms oxidise Fe^{2+}. The mechanism by which this occurs is uncertain and one suggestion is the following:

$$H + H^+ \longrightarrow H_2^+ \qquad (6.51)$$

$$H_2^+ + Fe^{2+} \longrightarrow Fe^{3+} + H_2 \qquad (6.65)$$

An alternative suggestion is that the H atom abstracts another H atom from a water molecule in the solvation shell of the ferrous ion:

$$Fe^{2+}(H_2O) + H \longrightarrow Fe^{3+}(OH^-) + H_2 \qquad (6.66)$$

One further suggestion has been that the H atom forms a hydride complex with the ferrous ion, the complex subsequently reacting with a hydrogen ion:

$$Fe^{2+} + H \longrightarrow (FeH)^{2+} \qquad (6.67)$$

$$(FeH)^{2+} + H^+ \longrightarrow Fe^{3+} + H_2 \qquad (6.68)$$

Whatever the explanation, experiment has established that in the absence of oxygen one H atom leads to the oxidation of one Fe^{2+}. In this case the radiolytic yield of Fe^{3+}, $G(Fe^{3+})_{vac}$, is given by

$$G(Fe^{3+})_{vac} = 2G_{H_2O_2} + G_{OH} + (G_H + e^-) \qquad (6.69)$$

Once again, equation (6.69) is valid only if the period of radiolysis is kept below that for which ferric ions will compete successfully for H atoms, as in reaction (6.63).

Comparison of equations (6.62) and (6.69) shows that the combined primary yields of H and e_{aq}^- may be deduced from a determination of the yield of ferric ions in the presence and absence of oxygen. Subtracting equation (6.69) from equation (6.62),

$$2(G_H + G_{e^-}) = G(Fe^{3+})_{O_2} - G(Fe^{3+})_{vac} \qquad (6.70)$$

In the above treatment the occurrence of HO_2 as a primary product has been neglected. This is justifiable with low–LET radiation but in the above case it would make no difference to equation (6.70). A term, $3G_{HO_2}$, would have to be added to both equations (6.62) and (6.69) and would thus cancel in the subtraction.

Ceric sulphate

In acid solutions of ceric sulphate, e_{aq}^- is once again converted to H by reaction (6.14). In the presence of oxygen, H is converted to HO_2 which then reduces ceric ions:

$$H + O_2 \longrightarrow HO_2 \qquad (6.34)$$

$$HO_2 + Ce^{4+} \longrightarrow Ce^{3+} + H^+ + O_2 \qquad (6.71)$$

I

The primary H_2O_2 also reduces ceric ions:

$$H_2O_2 + Ce^{4+} \longrightarrow Ce^{3+} + HO_2 + H^+ \qquad (6.72)$$

and the HO_2 so formed will reduce a further ceric ion by reaction (6.71). Once again, the contribution of primary HO_2 will be neglected. Each H_2O_2 molecule thus reduces two ceric ions and each H causes the reduction of one ceric ion.

The primary OH appear to be consumed in oxidising the cerous ions, which are the products of reactions (6.71) and (6.72):

$$OH + Ce^{3+} \longrightarrow Ce^{4+} + OH^- \qquad (6.73)$$

Any OH which reacts with the acid according to reaction (6.58) will still cause the oxidation of cerous ions which will react with the HSO_4 radicals formed. The over-all yield of cerous ions, $G(Ce^{3+})$, is thus given by

$$G(Ce^{3+}) = 2G_{H_2O_2} + (G_H + G_{e^-}) - G_{OH} \qquad (6.74)$$

It appears in this case that the product (cerous ions) of reactions occurring in the bulk of the solution can react immediately with a primary product (hydroxyl radicals). It might be thought that until the cerous ions build up to a concentration of about 10^{-6} mol dm^{-3} some of the hydroxyl radicals would escape scavenging by cerous ions and be free to take part in radical–radical reactions. It has been noted, however, that to obtain reproducible results with this system, a trace of cerous ion is required to be present in the ceric sulphate solutions. There is no doubt that the validity of equation (6.74) is confirmed by calculations of primary yields from other systems.

In the absence of oxygen the H atoms react directly with ceric ions:

$$H + Ce^{4+} \longrightarrow Ce^{3+} + H^+ \qquad (6.75)$$

Once again, each H leads to the reduction of one ceric ion and the radiolytic yield of cerous ions is still given by equation (6.74).

By subtracting equation (6.74) from equation (6.69) the yield of OH may be obtained:

$$2G_{OH} = G(Fe^{3+})_{vac} - G(Ce^{3+}) \qquad (6.76)$$

$G_H + G_{e^-}$ and G_{OH} having been calculated, these values may be substituted in any of equations (6.62), (6.69) or (6.74) in order to arrive at $G_{H_2O_2}$. G_{H_2} and G_{-H_2O} may then be obtained from the material balance equation (6.56). The value of $G_{H_2O_2}$ may be confirmed by measuring the yield of oxygen, $G(O_2)$, produced by the radiolysis of deoxygenated ceric sulphate solutions. In the absence of oxygen the reaction scheme may be written

$$e_{aq}^- + H_3O^+ \longrightarrow H + H_2O \tag{6.14}$$

$$H + Ce^{4+} \longrightarrow Ce^{3+} + H^+ \tag{6.75}$$

$$H_2O_2 + Ce^{4+} \longrightarrow Ce^{3+} + H^+ + HO_2 \tag{6.72}$$

$$HO_2 + Ce^{4+} \longrightarrow Ce^{3+} + H^+ + O_2 \tag{6.71}$$

$$OH + Ce^{3+} \longrightarrow Ce^{4+} + OH^- \tag{6.73}$$

From reactions (6.72) and (6.71) it can be seen that each molecule of H_2O_2 gives rise to one molecule of oxygen, so that

$$G(O_2) = G_{H_2O_2} \tag{6.77}$$

Ferrous sulphate–cupric sulphate

In the previous two systems the occurrence of HO_2 as a primary product has been neglected, as its formation is negligible with low-LET radiation and even with high-LET radiation G_{HO_2} is very small indeed. This quantity can be determined from the ferrous sulphate–cupric sulphate system, which, like the ceric sulphate system, is independent of the presence or absence of oxygen. The ferrous sulphate–cupric sulphate system has been studied in 5×10^{-3} mol dm^{-3} H_2SO_4 solutions, the cupric salt being maintained in tenfold excess of the ferrous salt.

In the presence of oxygen the reactions occurring are

$$e_{aq}^- + H_3O^+ \longrightarrow H + H_2O \tag{6.14}$$

$$H + O_2 \longrightarrow HO_2 \tag{6.34}$$

$$HO_2 + Cu^{2+} \longrightarrow Cu^+ + H^+ + O_2 \tag{6.78}$$

$$OH + Fe^{2+} \longrightarrow Fe^{3+} + OH^- \tag{6.57}$$

$$H_2O_2 + Fe^{2+} \longrightarrow Fe^{3+} + OH + OH^- \tag{6.60}$$

$$Cu^+ + Fe^{3+} \longrightarrow Cu^{2+} + Fe^{2+} \tag{6.79}$$

The excess of cupric sulphate ensures that all HO_2 is consumed by reaction (6.78) and has no chance of reacting with either ferrous ions or ferric ions according to reactions (6.39) and (6.64), respectively. From the above reaction scheme the radiolytic yield of ferric ions in the presence of excess cupric ions, $G(Fe^{3+})_{Cu^{2+}}$, is given by

$$G(Fe^{3+})_{Cu^{2+}} = 2G_{H_2O_2} + G_{OH} - (G_H + G_{e^-}) - G_{HO_2} \tag{6.80}$$

In the absence of oxygen, the H atoms reduce the cupric ions directly:

$$H + Cu^{2+} \longrightarrow Cu^+ + H^+ \tag{6.81}$$

rather than through HO_2 radicals. In this case the radiolytic yield of ferric ions will still be given by equation (6.80). If oxygen is initially absent from this system, it can be seen that any primary HO_2 produces oxygen by reaction (6.78). A determination of the amount of oxygen produced thus gives the primary yield of HO_2:

$$G(O_2)_{Fe^{2+}, \ Cu^{2+}} = G_{HO_2} \tag{6.82}$$

It will be realised that even if the system is studied with oxygen initially absent, some of this gas is soon generated by reaction (6.78). As G_{HO_2} is very small, the amount of oxygen generated will also be small; and as H atoms are fairly effectively scavenged by cupric ions (reaction 6.81), it is unlikely that reaction (6.34) occurs to any great extent. If any oxygen is consumed by this reaction it will, of course, be regenerated by reaction (6.78).

It will be recalled that in the discussion of the oxygenated ferrous sulphate system it was mentioned that the dose had to be kept below that where all the oxygen would be consumed. It has been suggested by Hart that, as the ferrous sulphate – cupric sulphate system is independent of oxygen, it could provide the basis of a dosimeter capable of measuring larger doses than the Fricke dosimeter.

NEUTRAL SOLUTIONS

Cationic solutes cannot be used for the determination of radical and molecular yields in neutral solutions, owing to the hydrolysis of the cations. These yields can be determined, however, by the combination of the results of several suitable systems, as has been done for the case of acid solutions above.

Bromide

It has already been pointed out that the molecular products which escape from the spurs in the radiolysis of pure water are attacked by the radical products. If a solute can be found which will scavenge the radicals but will not react with the molecular products, these products would be protected from attack by the radicals. Bromide ions will protect the molecular hydrogen in this way. In pure water the hydrogen is attacked by OH radicals, but these radicals can be scavenged by bromide ions. In deoxygenated solutions the reactions of bromide ions with the radical products are

$$OH + Br^- \longrightarrow Br + OH^- \tag{6.83}$$

$$Br + Br^- \longrightarrow Br_2^- \tag{6.84}$$

$$Br_2^- + H \longrightarrow HBr + Br^- \tag{6.85}$$

$$Br_2^- + e_{aq}^- \longrightarrow 2Br^- \tag{6.86}$$

It will be noticed that the Br atoms formed by reaction (6.83) associate with Br^- ions. This is really an equilibrium which also occurs in the radiolysis of chlorides and iodides, and in the bromide case it is greatly in favour of Br_2^- ions. As the bromide ions scavenge all the OH radicals, the molecular hydrogen is protected from attack by reaction (6.23) and

$$G(H_2) = G_{H_2} \tag{6.87}$$

This method is a direct method of determining G_{H_2} and can, of course, be used in acid solutions also.

In fact, there is a very slow reaction between Br_2^- and H_2 but its rate constant is about five orders of magnitude less than the rate constants of reactions (6.83) – (6.86). The reaction between Br_2^- and H_2 may thus be neglected.

Oxygen

In the same way that bromide ions can be used to scavenge OH, oxygen can be used to scavenge H and e_{aq}^-. A determination of the yield of H_2O_2 formed in the presence of oxygen can make a contribution to the determination of radical and molecular yields in neutral solution.

The H and e_{aq}^- are scavenged according to reactions (6.34) and (6.35), and the OH radicals can attack H_2O_2 by reaction (6.22):

$$H + O_2 \longrightarrow HO_2 \, (\to H^+ + O_2^-) \tag{6.34}$$

$$e_{aq}^- + O_2 \longrightarrow O_2^- \tag{6.35}$$

$$OH + H_2O_2 \longrightarrow H_2O + HO_2 \, (\to H^+ + O_2^-) \tag{6.22}$$

In neutral solution HO_2 dissociates to form O_2^- ions. The OH radicals will also attack these ions and the ions can combine to form H_2O_2:

$$OH + O_2^- \longrightarrow O_2 + OH^- \tag{6.88}$$

$$O_2^- + O_2^- \xrightarrow{(2H_2O)} H_2O_2 + O_2 + 2OH^- \tag{6.37}$$

Each H atom leads to the production of half a molecule of H_2O_2

and each OH reverses the effect of one H or e_{aq}^-. Under these circumstances,

$$G(H_2O_2)_{O_2} = G_{H_2O_2} + \tfrac{1}{2}(G_H + G_{e^-}) - \tfrac{1}{2}G_{OH} \qquad (6.89)$$

There is a complication in this system, in so far as some OH radicals could be consumed in attacking the molecular hydrogen. Sworski surmounted this problem by adding a trace of bromide, which scavenges all the OH radicals by reaction (6.83). The reaction between the resulting Br_2^- ions and H_2 is negligibly slow, but apparently Br_2^- reacts with H_2O_2 and O_2^- in the same way as OH.

By invoking the material balance equation, (6.56), equation (6.89) may be rearranged to give

$$G(H_2O_2)_{O_2} = 2G_{H_2O_2} - G_{H_2} \qquad (6.90)$$

If G_{H_2} is determined from deoxygenated bromide solutions, $G_{H_2O_2}$ may be deduced from a study of oxygenated bromide solutions.

Oxygen and hydrogen

In neutral solutions containing both oxygen and hydrogen it is found that the yield of H_2O_2 is independent of the concentrations of the dissolved gases over a considerable range. This must mean that the hydrogen gas is scavenging all OH radicals and the oxygen is scavenging all H and e_{aq}^-. The reactions occurring are thus

$$OH + H_2 \longrightarrow H + H_2O \qquad (6.23)$$

$$H + O_2 \longrightarrow HO_2 \, (\longrightarrow H^+ + O_2^-) \qquad (6.34)$$

$$e_{aq}^- + O_2 \longrightarrow O_2^- \qquad (6.35)$$

$$O_2^- + O_2^- \xrightarrow{\ (2H_2O)\ } H_2O_2 + O_2 + 2OH^- \qquad (6.37)$$

Each free radical thus leads to the production of half a molecule of H_2O_2 in addition to the molecular yield, and thus

$$G(H_2O_2)_{O_2,H_2} = G_{H_2O_2} + \tfrac{1}{2}(G_H + G_{e^-}) + \tfrac{1}{2}G_{OH} \qquad (6.91)$$

Subtraction of equation (6.89) from equation (6.91) gives

$$G(H_2O_2)_{O_2,H_2} - G(H_2O_2)_{O_2} = G_{OH} \qquad (6.92)$$

With $G_{H_2O_2}$ and G_{H_2} already determined, $G_H + G_{e^-}$ can be obtained from the material balance equation.

ALKALINE SOLUTIONS

Suitable systems for study in alkaline medium are difficult to find. Dainton and Watt have determined the radical and molecular yields from a study of solutions of ferrocyanide and ferricyanide in the presence and absence of nitrous oxide.

Ferrocyanide and ferricyanide

In alkaline solutions any primary H atoms are converted to e_{aq}^- by reaction (6.47):

$$H + OH^- \longrightarrow e_{aq}^- \tag{6.47}$$

Dainton and Watt have shown that the only reactions occurring in deoxygenated alkaline solutions of ferrocyanide, ferricyanide and nitrous oxide are

$$Fe(CN)_6^{3-} + e_{aq}^- \longrightarrow Fe(CN)_6^{4-} \tag{6.93}$$

$$N_2O + e_{aq}^- \longrightarrow N_2 + O^- \tag{6.94}$$

$$OH + Fe(CN)_6^{4-} \longrightarrow Fe(CN)_6^{3-} + OH^- \tag{6.95}$$

$$HO_2^- + 2Fe(CN)_6^{3-} \longrightarrow 2Fe(CN)_6^{4-} + H^+ + O_2 \tag{6.96}$$

It will be remembered that O^- and HO_2^- are the alkaline forms of OH and H_2O_2, respectively.

As all the primary H_2O_2 will be scavenged by reaction (6.96),

$$G_{H_2O_2} = G(O_2) \tag{6.97}$$

Similarly, all the OH radicals are scavenged by reaction (6.95) and the molecular hydrogen is protected, so

$$G_{H_2} = G(H_2) \tag{6.98}$$

In the absence of N_2O, all e_{aq}^- will react by reaction (6.93), and the yield of ferricyanide, $G(\text{ferri})$, under these conditions is given by

$$G(\text{ferri}) = G_{OH} - 2G_{H_2O_2} - (G_H + G_{e^-}) \tag{6.99}$$

Alternatively, if the concentration of N_2O and ferricyanide are such that all e_{aq}^- disappear by reaction (6.94), the yield of ferricyanide, $G(\text{ferri})_{N_2O}$, is given by

$$G(\text{ferri})_{N_2O} = G_{OH} - 2G_{H_2O_2} + (G_H + G_{e^-}) \tag{6.100}$$

Subtracting equation (6.99) from equation (6.100),

$$(G_H + G_{e^-}) = \tfrac{1}{2}[G(\text{ferri})_{N_2O} - G(\text{ferri})] \tag{6.101}$$

Alternatively,

$$(G_H + G_{e^-}) = G(N_2)_{N_2O} \qquad (6.102)$$

When we have obtained $G_{H_2O_2}$, G_{H_2} and $(G_H + G_{e^-})$ from the above relationships, G_{OH} may be obtained from the material balance equation.

OTHER METHODS

The methods of determination of the molecular and radical yields discussed above lead to individual values of $G_{H_2O_2}$, G_{H_2} and G_{OH} but the yield of reducing species appears only as the composite term $(G_H + G_{e^-})$. Attempts have been made to separate the components of this term by studying systems in which the solutes scavenge either e^-_{aq} or H. Some workers have used binary mixtures in which one component scavenges e^-_{aq} and the other component scavenges H. Examples of such mixtures are acetone and isopropanol, ferricyanide and formate, and bicarbonate and methanol, where the first components scavenge e^-_{aq} and the second components scavenge H atoms by hydrogen abstraction reactions. In other cases the total yield of reducing species has been determined with oxygenated H_2O_2 solutions and the yield of H atoms has been determined from the yield of hydrogen gas arising from the scavenging of H atoms by acetic acid. There is some discrepancy between the results obtained with organic and inorganic scavengers and those obtained with oxygenated H_2O_2 solutions.

The best methods of determining radical and molecular yields are probably the direct methods, where the yield of the species of interest can be measured directly rather than from the results of a series of reactions.

G_{H_2}

A direct method for measuring G_{H_2} has already been mentioned. This is the method in which the molecular hydrogen is protected from attack by bromide ions which scavenge the radical products which would otherwise attack the molecular hydrogen. The hydrogen is evolved and measured directly by gasometric or gas chromatographic techniques.

$G_{H_2O_2}$

This quantity has only been determined directly by measuring the yield of $H_2O_2^{18,18}$ formed from H_2O^{18}. The $H_2O_2^{18,18}$ must be protected from attack by the radicals and this is accomplished by adding $H_2O_2^{16,16}$ to the system. Solutions containing various concentrations of $H_2O_2^{16,16}$ are investigated and the G-values obtained for $H_2O_2^{18,18}$ are extrapolated to zero $H_2O_2^{16,16}$ concentration.

G_H

The yield of hydrogen atoms may be determined directly from the yield of HD obtained from solutions of substrates containing deuterium. Examples of such substrates are D_2, CD_3OH and $DCOO^-$. The primary hydrogen atoms take part in an abstraction reaction with the substrates to form HD.

G_{e^-}

The yield of e_{aq}^- may be measured directly by spectrophotometry coupled with pulse radiolysis. In such an experiment the pulse of energy generates sufficient electrons to give a measurable absorbance, which can be determined a very short time after the pulse before the hydrated electrons have time to disappear by reaction.

G_{OH}

There is no direct method available to measure G_{OH}. The best indirect methods depend on the reactions of OH radicals with substrates such as H_2, I^-, Br^- and $Fe(CN)_6^{4-}$.

With hydrogen in alkaline solutions the reactions are:

$$O^- + H_2 \longrightarrow H + OH^- \tag{6.45}$$

$$H + OH^- \longrightarrow e_{aq}^- \tag{6.47}$$

When the system is studied by pulse radiolysis, the yield of e_{aq}^- from reaction (6.47) can be measured spectrophotometrically.

With bromides and iodides the reactions are similar:

$$OH + Br^- \longrightarrow Br + OH^- \tag{6.83}$$

$$Br + Br^- \longrightarrow Br_2^- \tag{6.84}$$

and

$$OH + I^- \longrightarrow I + OH^- \tag{6.103}$$

$$I + I^- \longrightarrow I_2^- \tag{6.104}$$

In each case the species Br_2^- and I_2^- are reactive species and appear only transiently. They have characteristic absorption spectra, however, which give measurable absorbances when the systems are subjected to pulse radiolysis.

With ferrocyanide OH radicals give a stable product, unlike the above cases, where the first product is very reactive:

$$OH + Fe(CN)_6^{4-} \longrightarrow Fe(CN)_6^{3-} + OH^- \qquad (6.95)$$

In spite of the fact that ferricyanide is stable, it is best to study this system also by pulse radiolysis. In this way the ferricyanide which is initially formed by reaction (6.95) is measured directly before it can be attacked by H and e_{aq}^-. It thus gives as direct a measure as possible of the OH produced.

The above methods which involve pulse radiolysis are considered to be more reliable than the classical methods based on the over-all yields of oxidised products or on the material balance equation.

RESULTS OF RADICAL AND MOLECULAR YIELD DETERMINATIONS

As has been mentioned earlier, most work in radiation chemistry has been carried out with low-LET radiation such as γ-rays, X-rays and fast electrons. Most of the work with higher-LET radiation has been confined to acid solutions, so that it is best to restrict the consideration of the effect of LET on the primary product yields to acid solutions. Similarly, any consideration of the variation of primary yields with pH is more or less restricted to low-LET radiation.

EFFECT OF LET

Table 6.1 gives the most currently reliable yields of primary products for acid solutions of pH 0·5 arising from radiations of various LET. The LET values of the radiations increase from the lowest value for γ-rays to the highest value for 5·5 MeV α-particles.

Table 6.1 Variation of primary product yields with LET

Radiation	$G_{H_2O_2}$	G_{H_2}	G_{OH}	$(G_H + G_{e^-})$	G_{HO_2}	G_{-H_2O}
Co⁶⁰ γ-rays	0·8	0·45	2·95	3·65	0·008	4·55
H³ β-rays	1·0	0·6	2·1	2·9		4·1
18 MeV D⁺	1·03	0·7	1·75	2·4		3·85
8 MeV D⁺	1·2	1·05	1·45	1·7		3·85
32 MeV α	1·25	1·15	1·05	1·3		3·55
5·5 MeV α	1·34	1·57	0·5	0·6	0·11	3·5

The important effect of LET is to be seen in the relative magnitudes of the yields of the molecular products, compared with the yields of the radical products. With low-LET radiation, the yields of H_2 and H_2O_2 are much lower than the yields of OH, H and e_{aq}^-. As the LET of the radiation increases, the yields of the radical products decrease and the yields of the molecular products increase. The exception to this observation is G_{HO_2}, which is negligible with low-LET radiation but makes a small contribution with high-LET radiation. It was pointed out in the section on the mechanism of the radiolysis of water that the diffusion model would predict these effects. With low-LET radiation the radical concentrations will be low, and many radicals will escape into the bulk of the solution before undergoing radical–radical reactions in the spurs. As the molecular products are postulated to arise from radical–radical reactions, the yields of molecular products should be low under these conditions. With high-LET radiation, owing to the high radical concentration in the track, radical–radical reactions will occur to a great extent, leading to large yields of molecular products and correspondingly small yields of free radicals. As the HO_2 arises as a result of reaction (6.22),

$$OH + H_2O_2 \longrightarrow HO_2 + H_2O \qquad (6.22)$$

G_{HO_2} should be important only for high-LET radiation where the concentration of H_2O_2 in the track should be appreciable. With high-LET radiations, of course, in addition to reactions (6.18) and (6.19)

$$e_{aq}^- + H \longrightarrow H_2 + OH^- \qquad (6.18)$$

$$OH + H \longrightarrow H_2O \qquad (6.19)$$

occurring in the tracks, there is the possibility that some molecular hydrogen may be produced from the combination of H atoms:

$$H + H \longrightarrow H_2 \qquad (6.105)$$
$$k = 1 \cdot 0 \times 10^{10} \text{ mol}^{-1} \text{ dm}^3 \text{ s}^{-1}$$

As the yields of H and e_{aq}^- are much smaller at high values of LET it is to be expected that the influence of oxygen will be reduced at high LET. *Figure 6.2* illustrates this point; $G(Fe^{3+})$ for acid solutions of ferrous sulphate is plotted against LET for both aerated and anaerated solutions. It will be remembered that in de-aerated solution a hydrogen atom leads to the oxidation of one ferrous ion, but in aerated solution each hydrogen atom causes the oxidation of three ferrous ions through the formation of a perhydroxyl radical by reaction (6.34):

$$H + O_2 \longrightarrow HO_2 \qquad (6.34)$$

The difference between the values of $G(Fe^{3+})$ in aerated and de-aerated solutions is thus equal to $2(G_H + G_{e^-})$ (see equations 6.62 and 6.69). As the radical yields decrease with increasing LET, it is seen that the difference between $G(Fe^{3+})$ in aerated and de-aerated solutions becomes smaller. In addition to this effect, both values of $G(Fe^{3+})$ decrease with increasing LET because both G_{OH} and $(G_H + G_{e^-})$ decrease considerably while $G_{H_2O_2}$ increases only slightly.

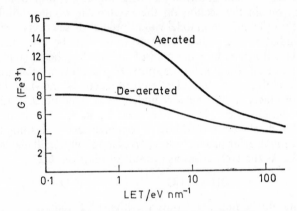

Figure 6.2.   Variation of G(*Fe*$^{3+}$) *with LET*

In *Table 6.1* G_{-H_2O} decreases slightly as the LET of the radiation increases. It must be remembered that G_{-H_2O} is the *net* yield of water decomposed to give radical and molecular products and is not equal to the total yield of water decomposed initially, as some of the radicals will combine to reform water. As the LET of the radiation increases, the radical concentration increases and thus more radicals will combine to reform water. Under these circumstances, the net yield of water decomposed will be less than for low-LET radiation.

There is no way of directly measuring the initial water decomposition but an approximation may be made on the basis of a reasonable assumption. The reactions by which the radicals form the molecular products are

$$e_{aq}^- + e_{aq}^- \longrightarrow H_2 + 2OH^- \tag{6.16}$$

$$OH + OH \longrightarrow H_2O_2 \tag{6.17}$$

$$e_{aq}^- + H \longrightarrow H_2 + OH^- \tag{6.18}$$

$$H + H \longrightarrow H_2 \tag{6.105}$$

The reactions by which the radicals reform water are

$$e_{aq}^- + OH \longrightarrow OH^- \qquad (6.15)$$

$$H + OH \longrightarrow H_2O \qquad (6.19)$$

If it is assumed that the combined probability of the first set of reactions occurring is about the same as for the latter reactions, then approximately equal amounts of H, OH and e_{aq}^- will react to reform water as react to form the molecular products H_2 and H_2O_2. If this assumption is true, then the yield of water initially decomposed, G_{-H_2O} (initial), will be given by

$$G_{-H_2O} \text{ (initial)} = G_{-H_2O} + G_{H_2O_2} + G_{H_2} \qquad (6.106)$$

Table 6.2 shows that G_{-H_2O} (initial) is approximately constant for radiations of various LETs. Another point brought out by *Table 6.2* is that G_{OH} decreases with increasing LET by the same factor as $(G_H + G_{e^-})$. Thus it can be seen that the ratio $(G_H + G_{e^-})/G_{OH}$ is very nearly constant for radiations of various LET values.

The above points about the constancy of G_{-H_2O} (initial) and $(G_H + G_{e^-})/G_{OH}$ imply that, at high LET, the radicals H, OH and e_{aq}^- are evenly distributed within the tracks, so that the chance of e_{aq}^- and H escaping from the tracks is in constant ratio to the chance of OH escaping from the tracks, even though the LET changes for the higher LET radiations. This conclusion is not in accord with the predictions of the diffusion model which, it will be remembered, suggests that the OH radicals are concentrated at the centre of the tracks while the hydrated electrons are more widely distributed. This picture would lead to the expectation that the chance of radical combination to give H_2 would be less than that to give H_2O, which would in turn be less than that to give H_2O_2.

Table 6.2 Effect of LET on the ratio of radical product yields and initial water decomposition yield

Radiation	$(G_H + G_{e^-})/G_{OH}$	G_{-H_2O} (initial)
Co⁶⁰ γ-rays	1·2	5·8
H³ β-rays	1·4	5·7
18 MeV D⁺	1·4	5·6
8 MeV D⁺	1·2	6·1
32 MeV α	1·2	6·0
5·5 MeV α	1·2	6·4

Studies of the H/D isotope effect on the formation of H_2 by the radiolysis of water with high-LET radiations have also indicated that the molecular hydrogen is not primarily formed by reaction (6.16) and that the hydrogen atoms do not originate from reaction (6.14). While the diffusion model seems to offer a good explanation of the dependence of radical and molecular yields on LET, there are still some doubts to be resolved.

EFFECT OF pH

Table 6.3 gives the radical and molecular yields for various pH values obtained with low-LET radiation. Most of the values quoted have been obtained by direct methods.

Table 6.3 Variation of radical and molecular yields with pH

pH	$G_{H_2O_2}$	G_{H_2}	G_{OH}	G_H	G_{e^-}	$(G_H + G_{e^-})$	G_{H_2O}
0–2	0·8	0·45	2·95	0·6	3·05	3·65	4·55
4–9	0·75	0·45	2·8	0·6	2·8	3·4	4·3
12	0·75	0·40	2·9	0·55	3·05	3·6	4·4

The yields of H_2, H_2O_2 and H are approximately independent of pH for low LET, while the yields of OH and e^-_{aq} are increased in both acid and alkaline solutions. One or two explanations of the variation in these radical yields with pH have been put forward in the past, before separate values of G_H and G_{e^-} were available, but there does not seem to be an entirely satisfactory theory available to explain the latest results given in *Table 6.3*.

EFFECT OF DOSE RATE

The usual magnitudes of dose rates in radiation chemistry lie between 10^{13} and 10^{17} eV cm^{-3} s^{-1}. Under these conditions, the tracks of the ionising particles are fairly widely separated and the radicals diffusing from one track will have reacted with the substrate before encountering any radicals diffusing from a neighbouring track. It is possible, experimentally, to achieve much higher dose rates with electron beams and, by restricting attention to this type of low-LET radiation, it will be appreciated that along the track of any one particle successive spurs will be some distance apart. Reactions between radicals from different spurs on the same track (intra-track reactions) have little chance of occurring. With very high dose rates, however, individual tracks will be situated quite close to one another and there is a good chance that radicals from a spur of one track may react with radicals from a spur of a different track (inter-track reactions) before being scavenged by solute. In this situation radical combination reactions to give the molecular products, H_2 and H_2O_2, should be encouraged, and the yields of radical products available for reaction with solute should decrease. The effect of an increase of dose rate with low-LET radiation should thus be analogous to the effect of increase of LET.

High dose rates can be achieved by using pulsed electron beams where the system is exposed to a high dose rate of up to 10^{25} eV cm^{-3} s^{-1} for a very short time of about 10^{-6} s. In this way the system receives only a small dose of energy of about 10^{19} eV cm^{-3} and excessive heating is avoided. This is, in fact, the technique of pulse radiolysis. *Figure 6.3* shows the results obtained for acid solutions of ferrous sulphate as a function of dose rate, and a comparison of *Figures 6.2* and *6.3* shows that the effects of increased dose rate are very similar to those of increased LET.

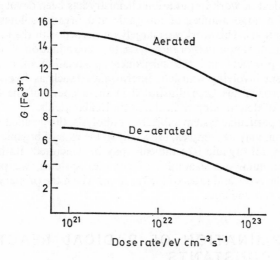

Figure 6.3. Effect of dose rate on $G(Fe^{3+})$

It is also found that at high dose rates the *G*-values become more sensitive to substrate concentration. This can be understood when it is remembered that *G*-values vary with concentration for high-LET radiation.

The decrease in the radiolytic yield of ferric ions with increasing dose rate can be understood qualitatively in terms of the diffusion model. *Figure 6.3*, however, shows that $G(Fe^{3+})$ decreases from dose rates of about 10^{21} eV cm^{-3} s^{-1} upwards. Kinetic calculations, made by Kupperman, based on the diffusion model, predict that effects due to the overlap of tracks at high dose rates should not become noticeable until the dose rate exceeds about 10^{24} eV cm^{-3} s^{-1}. Thus, although the diffusion model fits the observations qualitatively, the quantitative aspects leave something to be desired.

Some more work with pulse radiolysis at high dose rates shows deviations from the behaviour predicted by the diffusion model. These discrepancies can be explained by postulating the existence of H_3O radicals as precursors of hydrogen atoms; this point will be considered again later in the chapter.

RADIOLYSIS OF AQUEOUS SOLUTIONS AT LOW CONCENTRATIONS

A great deal of work in radiation chemistry has been devoted to the study of a large number of inorganic and organic solutes at low concentrations. This work was largely concerned with the identification of the final products of radiolysis, the determination of the yields of such products and the subsequent postulation of a reaction mechanism involving transient intermediates such as free radicals. This approach has been illustrated in the account of the classical methods of determining radical and molecular yields.

In any particular system subject to radiolysis, the detailed reaction mechanism may be complex, especially in the case of organic solutes, where several organic free radicals may be generated. Rather than give here numerous examples of complex systems, we prefer to discuss the types and rates of the reactions which the primary radical products can undergo.

DETERMINATION OF RADICAL REACTION RATE CONSTANTS

In the earlier days of radiation chemistry it was not possible to determine the rate constant of a reaction involving a primary radical. The final products of radiolysis were often the result of several reactions and any particular step of a mechanism involving a radical could not be directly followed. Such steps were postulated only on the basis of the final products. It was, however, possible in many cases to determine the relative values of the rate constants for the reactions of a particular radical with two different solutes.

If a primary radical, R, reacts with each of two solutes, A and B according to

$$R + A \longrightarrow P \qquad (6.107)$$

$$R + B \longrightarrow Q \qquad (6.108)$$

and if these are the only reactions in which the radical takes part, then every radical produced must disappear by reaction (6.107) or reaction (6.108). The fraction of radicals reacting with solute A to produce P is thus $_{107}v/(v_{107} + v_{108})$, where v_{107} and v_{108} are the velocities of reactions (6.107) and (6.108), respectively. The radiolytic yield of P, $G(P)$, is thus given by

$$G(P) = G_R \frac{v_{107}}{v_{107} + v_{108}}$$

or

$$G(P) = G_R \frac{k_{107}[R][A]}{k_{107}[R][A] + k_{108}[R][B]} \qquad (6.109)$$

Rearranging equation (6.109) gives

$$G(P) = G_R \frac{1}{1 + k_{108}[B]/k_{107}[A]} \qquad (6.110)$$

From a determination of $G(P)$ and a knowledge of G_R, [A] and [B], it is thus possible to obtain a value of k_{108}/k_{107}.

With the advent of the technique of pulse radiolysis it became possible to determine the rate constants of individual reactions involving radicals, by observing the rate of disappearance of the radicals directly. This type of information, coupled with the previous results for relative rate constants, has made possible the amassing of a great number of data on the rate constants of the reactions of hydroxyl radicals, hydrogen atoms and hydrated electrons with various solutes. This knowledge, together with that of the types of reaction undergone by primary radicals probably provides the best summary of the radiolytic behaviour of many aqueous solutions.

REACTIONS OF HYDRATED ELECTRONS

Hydrated electrons generally undergo addition reactions with substrates, the electron being accommodated in a vacant orbital of the substrate. The species so formed may be a stable product, as exemplified by reaction (6.93):

$$e_{aq}^- + Fe(CN)_6^{3-} \longrightarrow Fe(CN)_6^{4-} \qquad (6.93)$$

K

Alternatively, the adduct may be highly reactive and have only a short lifetime. There are four ways in which such an intermediate may react.

REACTION WITH WATER

This type of reaction is illustrated in the reaction of hydrated electrons with carbonyl compounds to form the corresponding alcohols. With acetone the first stage in the reaction is the formation of the intermediate:

$$e_{aq}^- + (CH_3)_2CO \longrightarrow (CH_3)_2CO^- \qquad (6.111)$$

The intermediate then reacts with water to form a free radical and a hydroxide ion:

$$(CH_3)_2CO^- + H_2O \longrightarrow (CH_3)_2COH + OH^- \qquad (6.112)$$

The main reaction of the free radicals is a disproportionation reaction to form isopropanol and acetone:

$$2(CH_3)_2COH \longrightarrow (CH_3)_2CO + (CH_3)_2CHOH \qquad (6.113)$$

DISSOCIATION

The intermediate may split into two fragments, one being an anion. This is illustrated by the reaction involving methyl chloride where the intermediate splits into a chloride ion and a methyl radical:

$$e_{aq}^- + CH_3Cl \longrightarrow CH_3Cl^- \qquad (6.114)$$

$$CH_3Cl^- \longrightarrow CH_3 + Cl^- \qquad (6.115)$$

The methyl radicals will, of course, undergo further reactions which will depend upon the species present in the system. The above reactions are analogous to the production of chloride ions from monochloracetic acid by reaction (6.50). (See also reaction 6.94.)

ELECTRON TRANSFER

Sometimes the intermediate may transfer an electron to an acceptor. For example, hydrated electrons can add to carbon dioxide and the intermediate so formed can transfer an electron to a methyl chloride molecule:

$$e_{aq}^- + CO_2 \longrightarrow CO_2^- \tag{6.116}$$

$$CO_2^- + CH_3Cl \longrightarrow CO_2 + CH_3Cl^- \tag{6.117}$$

The species CH_3Cl^- will then dissociate according to reaction (6.115).

DISPROPORTIONATION

If the electron acceptor in an electron transfer reaction is another molecule of the same intermediate, the reaction may be called a disproportionation reaction. For example, hydrated electrons will add to oxygen by reaction (6.35) to form O_2^-:

$$e_{aq}^- + O_2 \longrightarrow O_2^- \tag{6.35}$$

The disproportionation of O_2^- has already been illustrated by reaction (6.37):

$$O_2^- + O_2^- \xrightarrow{\text{(2H}_2\text{O)}} H_2O_2 + O_2 + 2OH^- \tag{6.37}$$

although the disproportionation might be better shown by writing reaction (6.37) in the form

$$O_2^- + O_2^- \xrightarrow{\text{(2H}^+\text{)}} O_2 + H_2O_2 \tag{6.118}$$

The rates of reactions involving hydrated electrons depend upon the availability of a vacant orbital on the substrate, and the rate constants of most of these reactions lie in the range 10^{10}–10^5 mol^{-1} dm^3 s^{-1}.

REACTIONS OF HYDROGEN ATOMS

Hydrogen atoms can undergo several types of reactions, one of which is their conversion to hydrated electrons in alkaline solutions by reaction (6.47):

$$H + OH^- \longrightarrow e_{aq}^- \tag{6.47}$$

Another reaction which stands alone is the reaction of hydrogen atoms with hydrogen peroxide:

$$H + H_2O_2 \longrightarrow OH + H_2O \tag{6.21}$$

Most of the other reactions of hydrogen atoms can be classed as abstraction reactions, addition reactions, charge transfer reactions or radical combination reactions, and these will be illustrated below.

ABSTRACTION REACTIONS

Probably the commonest type of abstraction reaction is that in which another hydrogen atom is abstracted. For example, with ethanol and acetic acid the reactions are

$$H + CH_3CH_2OH \longrightarrow H_2 + CH_2CHOH \qquad (6.119)$$

$$H + CH_3COOH \longrightarrow H_2 + CH_2COOH \qquad (6.120)$$

The resultant organic radicals will, of course, undergo further reactions.

Abstraction reactions are not limited to the abstraction of hydrogen. Halogen atoms can be abstracted from halogen molecules:

$$H + Br_2 \longrightarrow HBr + Br \qquad (6.121)$$

Amino acids can similarly be deaminated:

$$H + CH_2(NH_2)COOH \longrightarrow CH_2COOH + NH_3 \qquad (6.122)$$

ADDITION REACTIONS

Hydrogen atoms can add to unsaturated compounds and also to benzene:

$$H + PhH \longrightarrow PhH_2 \qquad (6.123)$$

In the radiolysis of benzene the radical PhH_2 is thought to be responsible for a polymeric product which forms.

In the case of unsaturated compounds the hydrogen atom adds on at the double bond generating an organic free radical:

$$RCH = CH_2 + H \longrightarrow RCHCH_3 \qquad (6.124)$$

These free radicals can initiate polymerisation of the unsaturated substrate, which can lead to products of high relative molecular masses, if the concentration of substrate is high compared with the concentration of primary radicals.

CHARGE TRANSFER REACTIONS

Hydrogen atoms can act as simple reducing agents in charge transfer reactions. Such reactions are illustrated by reactions (6.63), (6.75) and (6.81):

$$H + Fe^{3+} \longrightarrow Fe^{2+} + H^+ \qquad (6.63)$$

$$H + Ce^{4+} \longrightarrow Ce^{3+} + H^+ \qquad (6.75)$$

$$H + Cu^{2+} \longrightarrow Cu^+ + H^+ \qquad (6.81)$$

RADICAL COMBINATION REACTIONS

In common with other free radicals, hydrogen atoms can combine with free radicals in reactions such as (6.19), (6.105) and (6.34):

$$H + OH \longrightarrow H_2O \qquad (6.19)$$

$$H + H \longrightarrow H_2 \qquad (6.105)$$

$$H + O_2 \longrightarrow HO_2 \qquad (6.34)$$

In this last case it will be remembered that oxygen may be considered as a diradical because its ground state is a triplet.

The rates of radical combination reactions are usually quite high, of the order of 10^{10} mol^{-1} dm^3 s^{-1}. The rates of hydrogen abstraction reactions from aliphatic compounds vary and depend upon the strength of the carbon–hydrogen bond and upon the stability of the free radical which results from hydrogen abstraction. Halogen atoms can also be abstracted from aliphatic compounds; the rates of these reactions are governed by considerations similar to those which apply to hydrogen abstraction reactions. The rates of addition reactions are usually one or two orders of magnitude faster than hydrogen abstraction reactions.

REACTIONS OF HYDROXYL RADICALS

The reactions of hydroxyl radicals, like those of hydrogen atoms, can be described as abstraction reactions, addition reactions, electron transfer reactions and radical combination reactions. It must also be remembered that the hydroxyl radical undergoes dissociation according to reaction (6.44):

$$OH \rightleftharpoons H^+ + O^- \qquad (6.44)$$

so that in strongly alkaline solutions it is present as O^-.

ABSTRACTION REACTIONS

Once again, the commonest abstraction reactions are hydrogen abstraction reactions:

$$OH + CH_3CH_2OH \longrightarrow H_2O + CH_3CHOH \qquad (6.125)$$

$$OH + H_2 \longrightarrow H_2O + H \qquad (6.23)$$

$$OH + H_2O_2 \longrightarrow H_2O + HO_2 \qquad (6.22)$$

ADDITION REACTIONS

Like hydrogen atoms, hydroxyl radicals can add to benzene to form the intermediate PhHOH, which has been detected in pulse radiolysis studies:

$$OH + PhH \longrightarrow PhHOH \tag{6.126}$$

Once again, with unsaturated substances, a hydroxyl radical will add at the double bond to generate an organic free radical, which may initiate polymerisation in the system:

$$OH + RCH{=}CH_2 \longrightarrow RCHCH_2OH \tag{6.127}$$

ELECTRON TRANSFER REACTIONS

The hydroxyl radical may behave as a simple oxidising agent in electron transfer reactions such as those exemplified by reactions (6.95) and (6.103):

$$OH + Fe(CN)_6^{4-} \longrightarrow Fe(CN)_6^{3-} + OH^- \tag{6.95}$$

$$OH + I^- \longrightarrow I + OH^- \tag{6.103}$$

RADICAL COMBINATION REACTIONS

This type of reaction is illustrated by reactions (6.19) and (6.17):

$$OH + H \longrightarrow H_2O \tag{6.19}$$

$$OH + OH \longrightarrow H_2O_2 \tag{6.17}$$

As with hydrogen atoms, the addition reactions of hydroxyl radicals are faster than their hydrogen abstraction reactions. The rates of hydrogen abstraction by hydroxyl radicals are, however, considerably greater than the rates at which hydrogen is abstracted by hydrogen atoms. The reason for this difference is thought to be in the difference between the strengths of the H—H and HO—H bonds.

DEFECTS OF THE DIFFUSION MODEL OF WATER RADIOLYSIS

Although the diffusion model of radiolysis as presented earlier in this chapter is still the generally accepted model of the mechanism of radiolysis, it is realised that it has some deficiencies. One of these has already been pointed out in considering the effect of dose rate on the radical and molecular yields. Without going into the rather complicated mathematics of the diffusion kinetics, it was sufficient to say that the qualitative effect of dose rate accords with the diffusion model, but the model predicts quantitatively that a dose rate of two or three orders of magnitude greater than that used should be required to produce the observed effects. There are other points on which the diffusion model fails. When solutions of concentrations greater than about 10^{-3} mol dm^{-3} are studied, the yields of molecular products are found to decrease with increasing substrate concentration. The observed molecular yields, G, can be related to the substrate concentration, c, by

$$G = G^\infty - ac^{1/n} \qquad (6.128)$$

where G^∞ is the molecular yield at infinite dilution. The constant a depends on the nature of the solute and n has the value 3 for most systems studied at low LET, this value seeming to be smaller at higher LET. Although a is a constant for a particular solute in the concentration range 10^{-4}–10^{-1} mol dm^{-3}, at concentrations exceeding 0·5 mol dm^{-3} the value of a decreases, but more slowly than predicted by the diffusion model calculations. In addition, although $n = 3$ for most systems at various low-LET values, according to the diffusion model n should only have a value of 3 for the particular case of 1 MeV electrons.

The diffusion model postulates that the yield of molecular hydrogen arises largely from reaction (6.16):

$$e_{aq}^- + e_{aq}^- \longrightarrow H_2 + 2OH^- \qquad (6.16)$$

but some doubt has already been cast on this idea as a result of the effects of LET on molecular yields. Other evidence comes from a study of nitrate solutions. Nitrate ions react with hydrated electrons and hydrogen atoms, the rate constants of these reactions being about 10^{10} and 10^7 mol^{-1} dm^3 s^{-1}, respectively. The effect of nitrate ions on G_{H_2} is, however, independent of pH even down to strongly acid solutions. This is unexpected if hydrated electrons are being converted to hydrogen atoms in the spurs by reaction (6.14):

$$e_{aq}^- + H_3O^+ \longrightarrow H + H_2O \qquad (6.14)$$

If this were the case, nitrate ions which scavenge hydrated electrons more efficiently than they scavenge hydrogen atoms should cause a greater decrease in G_{H_2} in neutral solutions if molecular hydrogen is formed by reaction (6.16).

The effect of nitrous oxide on G_{H_2} is also independent of pH, and in this case the difference between the rate constants of the reactions of N_2O with e_{aq}^- and H is even greater than for nitrate ions. In fact, $k(e_{aq}^- + N_2O) = 8.7 \times 10^9 \, mol^{-1} \, dm^3 \, s^{-1}$ and $k(H + N_2O) = 2.2 \times 10^5 \, mol^{-1} \, dm^3 \, s^{-1}$.

In addition to the above evidence, there are various substrates, NO_3^-, Co^{2+}, Ni^{2+}, Tl^+, $Co(NH_3)_6^{3+}$, $Fe(CN)_6^{3-}$, which affect G_{H_2} to different extents while the rate constants for their reactions with e_{aq}^- are all very similar. The ion $Co(NH_3)_6^{3+}$ is very effective in reducing G_{H_2} but does not react rapidly with H, the appropriate rate constant being about $10^6 \, mol^{-1} \, dm^3 \, s^{-1}$. On the other hand, Tl^+ has no effect on G_{H_2} although its reaction with H has a rate constant of about $10^8 \, mol^{-1} \, dm^3 \, s^{-1}$. This last piece of evidence also seems to exclude reaction (6.105) as being responsible for the yield of molecular hydrogen:

$$H + H \longrightarrow H_2 \qquad (6.105)$$

In the same way that evidence can be adduced for dismissing reactions (6.16), (6.105) and even (6.18) as the route for molecular hydrogen, there is also evidence which contradicts the postulate that OH radicals are the precursors of molecular H_2O_2. Thus the ions Tl^+, $Fe(CN)_6^{4-}$ and I^- affect $G_{H_2O_2}$ to different extents, while the rate constants of their reactions with OH are all about $10^{10} \, mol^{-1} \, dm^3 \, s^{-1}$. Similarly, I^- and Br^- affect $G_{H_2O_2}$ to the same extent, while their rate constants for reaction with OH are vastly different.

There is thus a fair amount of evidence, of which the above is only a selection, indicating that e_{aq}^- and H are not the main precursors of H_2 and that OH is not the main precursor of H_2O_2. There is some support for these conclusions from the differences in the yields of primary products observed in D_2O as opposed to H_2O. This evidence is less convincing, however, as the argument rests on a somewhat tentative assumption.

MODIFICATIONS TO THE DIFFUSION MODEL

Since the classical diffusion model was first postulated, it has undergone many modifications. These modifications have been necessary as new experimental facts, such as the existence of the hydrated electron, came to light. As the current version of the diffusion model has some deficiencies, new modifications are still being suggested, and the purpose of this section is to summarise them.

The main problems which must be considered are the formation of H_2 and H_2O_2 by mechanisms in which H or e_{aq}^- and OH are not the precursors. Anbar has recently brought together some suggestions from various sources to present an interpretation of the mechanism of radiolysis of water, which, although not perfect, overcomes many of the outstanding problems of the diffusion model.

To account for at least some part of the molecular H_2O_2 yield, the reactions of the OH^+ ion have been invoked. It will be remembered that this ion has been observed in the mass spectrometry of water vapour but its reactions have, in the past, been assumed to give only H_2O^+ and OH. The formation of the OH^+ ion is given by reaction (6.2):

$$H_2O \longrightarrow\!\!\!\backsim\!\!\backsim\!\!\backsim\!\!\backsim\longrightarrow OH^+ + H + e^- \qquad (6.2)$$

It is now suggested that the OH^+ subsequently reacts according to

$$OH^+ + H_2O \longrightarrow H_3O^+ + O \qquad (6.129)$$

$$O + H_2O \longrightarrow H_2O_2 \qquad (6.130)$$

This scheme provides a route for the formation of H_2O_2 in which the precursor is O and not OH. Without going into details, we can say that the scheme accounts for many observations which would be anomalous on the basis of the diffusion model. The electron produced in reaction (6.2) is considered to become hydrated.

Some of the reactions of the classical model are included, such as

$$H_2O \longrightarrow\!\!\!\backsim\!\!\backsim\!\!\backsim\!\!\backsim\longrightarrow H_2O^+ + e^- \qquad (6.8)$$

$$H_2O^+ + H_2O \longrightarrow H_3O^+ + OH \qquad (6.11)$$

$$e^- + aq \longrightarrow e_{aq}^- \qquad (6.13)$$

There have been two suggestions to account for the formation of molecular H_2 by mechanisms not involving e_{aq}^- or H as precursors. Platzman has suggested that subexcitation electrons are captured by water molecules to form H^-, which subsequently reacts with water to form H_2:

$$e_{sub}^- + H_2O \longrightarrow H^- + OH \qquad (6.131)$$

$$H^- + H_2O \longrightarrow H_2 + OH^- \qquad (6.132)$$

As the subexcitation electrons will be distributed throughout the medium, this scheme implies that some, at least, of the molecular hydrogen is not formed in spurs.

An alternative suggestion, put forward by Hamill, is that some electrons are reduced to thermal energies before becoming hydrated. These thermal non-hydrated electrons can move very rapidly through the medium and react with H_2O^+ before this ion can react with water. The result is a highly excited water molecule which can dissociate to give 'hot' hydrogen atoms, which then react with water to form H_2:

$$H_2O^+ + e^- \longrightarrow H_2O^{**} \qquad (6.133)$$

$$H_2O^{**} \longrightarrow H^* + OH \qquad (6.134)$$

$$H^* + H_2O \longrightarrow H_2 + OH \qquad (6.135)$$

Once again, if the non-hydrated electrons can move rapidly through the medium, the molecular hydrogen may be formed in the bulk of the solution. A difficulty associated with this scheme is the short half-life of H_2O^+ in water. It may be that there is little chance of a non-hydrated electron reacting with H_2O^+ if this latter species disappears rapidly by reaction (6.11). This problem may be overcome by postulating that the non-hydrated electrons react with hydrogen ions to form H_3O. This species has been identified in the gas phase and there is some evidence for its existence in the liquid phase. The objection could thus be overcome by the following scheme:

$$H_3O^+ + e^- \longrightarrow H_3O \qquad (6.136)$$

$$H_3O \longrightarrow H^* + H_2O \qquad (6.137)$$

$$H^* + H_2O \longrightarrow H_2 + OH \qquad (6.135)$$

In either case differences in the reactivities of hydrated and non-hydrated electrons could account for anomalies in radiation chemistry which are usually attributed to the occurrence of spurs.

It should be pointed out that the reactions proposed above which lead to the molecular products involve a radical and a solvent molecule rather than two radicals as proposed in the classical diffusion model involving reactions in spurs. These modifications do not exclude spurs but it has been suggested that their contribution to the over-all process at low LET is small, most radiolytic events occurring homogeneously throughout the medium. Some current work on pulse radiolysis is concerned with the search for spurs but no direct evidence for their existence has as yet been found.

With high-LET radiation the local density of radicals will be high and in this case radical–radical reactions are probably predominant, although, as pointed out in the section on the effect of LET on radical and molecular yields, it seems that in this case the distribution of radicals within the tracks is homogeneous.

7

Gases and solids

The radiation chemistry of gases and solids will differ from that of water and aqueous solutions because of differences in the electron densities of the absorbers and variations in the related freedom of the excited and ionised species.

GASES

Gases, in particular, will differ from the condensed phases, as the diffusion of the transient species will be faster. They will have longer lifetimes and therefore the track effects will be less important. Radiation yields will be less dependent on the LET of the radiation than they were in water. The active species becomes more uniformly distributed in gases than in liquids or solids. Another feature in the irradiation of the gaseous state is that the ionised species can be determined by means of the mass spectrometer. This machine will also show up subsequent ion–molecule reactions as long as they take place within the time of flight of the spectrometer ($c.10^{-6}$s). A quantity which is more accurately known in the gas phase is the value of W, the energy dissipated in the formation of an ion pair. From this, and the saturation ionisation current, the number of ions and electrons produced as a result of irradiation can be obtained. Reactions which occur in irradiated gases are similar to those which take place in electrical discharges and in the ionosphere. The main gaseous reactions which have been investigated with ionising radiations are listed in *Table 7.1*.

Table 7.1 Radiation chemical reactions of gases

Gases irradiated	Main products
para-H_2	ortho-H_2
$H_2 + D_2$	HD
$H_2 + Cl_2$	HCl
$H_2O(g)$	H_2, H_2O_2
O_2	O_3
$N_2 + O_2$	N_2O, NO, NO_2, HNO_3
N_2O	N_2, NO
NH_3	N_2, N_2H_4, H_2
CO_2	CO, C_2O, C_3O_2
CO	CO_2, C_2O, C_3O_2, C(graphite)
CH_4	H_2, C_2H_6, C_3H_8
C_2H_6	H_2, C_4H_{10}, CH_4, C_3H_8, C_5H_{10}
C_2H_4	H_2, C_4H_{10}, CH_4, C_3H_8, C_5H_8
C_2H_2	C_6H_6, 'cuprene' polymer

Some of these reactions will now be discussed in detail.

HYDROGEN AND DEUTERIUM

The second and third systems shown in *Table 7.1* involve chain reactions. They have initiation, propagation and termination steps. The initiation step consists of the formation of an active species derived from one of the reactants, which can occur by both the dissociation of an excited molecular state and the reactions of the ions formed in the ionisation process. In the production of HD the initiation steps are

$$H_2 \longrightarrow\!\!\!\sim\!\!\!\sim\!\!\!\rightarrow H_2^* \longrightarrow 2H \qquad \text{excitation and dissociation} \qquad (7.1)$$

$$H_2 \longrightarrow\!\!\!\sim\!\!\!\sim\!\!\!\rightarrow H_2^+ + e^- \qquad \text{ionisation} \qquad (7.2)$$

$$H_2^+ + H_2 \longrightarrow H_3^+ + H \qquad \text{ion–molecule reaction} \qquad (7.3)$$

$$H^+ + e^- \longrightarrow 3H \qquad \text{neutralisation–decomposition} \qquad (7.4)$$

Once the active species H_3^+ and H have been formed, the chain propagation steps may occur by both a free radical mechanism and an ionic mechanism. The former reactions may be represented,

$$H + D_2 \longrightarrow D + HD \qquad (7.5)$$

$$D + H_2 \longrightarrow H + HD \qquad (7.6)$$

The ionic propagation may be written

$$H_3^+ + D_2 \longrightarrow HD_2^+ + H_2 \qquad (7.7)$$

$$HD_2^+ + H_2 \longrightarrow H_2D^+ + HD \qquad (7.8)$$

$$H_2D^+ + D_2 \longrightarrow HD_2^+ + HD \qquad (7.9)$$

The HD is formed with a G-value of 6×10^4 molecules/100 eV. Normally termination of the radical chain takes place on the surface of the reaction vessel or in the gas phase through a three-body collision, and ionic propagation is terminated by neutralisation by electrons. Lower yields than the above are observed when trace impurities such as oxygen are present. These not only terminate the reaction, but also lower the radiation yield by removal of the chain carrier:

$$H + O_2 \longrightarrow HO_2 \longrightarrow \tfrac{1}{2}H_2O_2 + \tfrac{1}{2}O_2 \qquad (7.10)$$

On the other hand, the ionic propagation chain is affected by the introduction of certain noble gases with ionisation energies lower than those of the possible ionic chain carriers, H^+, H_2^+, H_3^+. This has the effect of removing some of the active species before propagation:

$$H_3^+ + Xe \longrightarrow XeH^+ + H_2 \qquad (7.11)$$

HYDROGEN AND CHLORINE

The formation of hydrogen chloride from hydrogen and chlorine is another chain reaction which can be initiated by ionising radiation through the ionisation and excitation of chlorine or hydrogen. The latter has been discussed above and the ionisation and excitation of chlorine may be represented

$$Cl_2 \longrightarrow\!\!\!\sim\!\!\!\sim\!\!\!\sim\!\!\!\sim 2Cl \qquad (7.12)$$

$$Cl_2 \longrightarrow\!\!\!\sim\!\!\!\sim\!\!\!\sim\!\!\!\sim Cl_2^+ + e^- \qquad (7.13)$$

$$Cl_2^+ + Cl_2 \longrightarrow Cl^+ + Cl \qquad (7.14)$$

$$Cl_3^+ + e^- \longrightarrow Cl_2 + Cl \qquad (7.15)$$

The chlorine or hydrogen atoms then take part in the well-known propagation reaction

$$Cl + H_2 \longrightarrow H + HCl \qquad (7.16)$$

$$H + Cl_2 \longrightarrow Cl + HCl \qquad (7.17)$$

which is finally terminated by

$$2Cl + M \longrightarrow Cl_2 + M \qquad (7.18)$$

this reaction occurring in the gas phase or on the walls of the vessel, or both.

WATER VAPOUR

In the absence of a chain reaction, more normal radiation yields ($G \approx 10$) are observed in gases. In the case of water vapour, however, the yield is still larger than that found in liquid water. Whereas in the latter the value of $G-_{H_2O}$ is approximately 4·5, in the vapour phase it comes to 11·7 molecules/100 eV. This indicates that in the vapour phase recombination and de-excitation processes are not favoured because of the low density of the phase.

OXYGEN

When reactions leading to the destruction of the radiation product take place at the same time, the radiation yields are liable to be very small indeed ($G \approx 0.4$). The formation of ozone from oxygen is an example of this. In this case great variations in $G(O_3)$ are observed, because the radiation product can decompose through a chain reaction. In the first place, the reactions leading to ozone formation can be ascribed to the following:

$$O_2 \rightsquigarrow O_2^+ + e^- \tag{7.19}$$

$$O_2 \rightsquigarrow O_2^* \longrightarrow 2O \tag{7.20}$$

$$O_2 + e^- \longrightarrow O_2^- \tag{7.21}$$

$$O_2^* + O_2 \tag{7.22}$$

$$O_2^- + O_2^+ \longrightarrow O_3 + O \tag{7.23}$$

$$2O + O_2 \tag{7.24}$$

$$O + O_2 + M \longrightarrow O_3 + M \tag{7.25}$$

The oxygen atoms involved in the above reactions may or may not be in an excited state.

Decomposition of the radiation product O_3 then occurs by

$$O_3 + O \longrightarrow 2O_2 \qquad (7.26)$$

or by

$$O_3 + O \longrightarrow 2O + O_2 \qquad (7.27)$$

If reaction (7.27) occurs, the decomposition of ozone becomes a chain reaction and it is this which causes the wide variation (2–124 molecules/100 eV) in the values of $G(O_3)$ reported in the literature.

The normal boiling points of O_3, O_2 and N_2 are 161 K, 90 K and 77 K, respectively. If, therefore, liquid nitrogen traps are used during an irradiation process and air is also present, liquid ozone and oxygen will condense in the trap. Great care should then be taken when warming the trap to room temperature, because organic matter coming into contact with liquid ozone is liable to explode.

NITROGEN AND OXYGEN

The fixation of nitrogen has been widely investigated. This can be brought about by irradiating the gas in the presence of oxygen, where it will form various oxides of nitrogen. These, in turn, decompose back to the reactant gases and after a time an equilibrium is established. The work described in the previous section showed the effect of irradiation on oxygen, when atomic oxygen and ozone were formed. In the present case the reactions of nitrogen must also be considered and the following reaction scheme has been postulated:

$$N_2 \longrightarrow\!\!\!\sim\!\!\!\sim\!\!\!\sim\!\!\longrightarrow N_2^+ + e^- \qquad (7.28)$$

$$N_2 \longrightarrow\!\!\!\sim\!\!\!\sim\!\!\!\sim\!\!\longrightarrow 2N \qquad (7.29)$$

$$N_2^+ + O_2 \longrightarrow NO^+ + NO \qquad (7.30)$$

$$NO^+ + e^- \longrightarrow N + O \longrightarrow NO + h\nu \qquad (7.31)$$

$$N + O_2 \longrightarrow O + NO \qquad (7.32)$$

$$N + O_2M \longrightarrow NO_2M \qquad (7.33)$$

$$2NO + O_2 \longrightarrow 2NO_2 \qquad (7.34)$$

$$N + NO_2 \longrightarrow O + N_2O \qquad (7.35)$$

$$O + O_2 + M \longrightarrow O_3 + M \qquad (7.25)$$

In the presence of water vapour the oxidised products, NO, NO_2 and N_2O react further to form nitrous and nitric acid.

NITROUS OXIDE

If nitrous oxide is irradiated alone, the products are nitrogen, oxygen and nitrogen dioxide. At pressures greater than 50 Torr ($6 \cdot 7$ kN m^{-2}), the yields of nitrogen and nitrogen dioxide are consistently linear functions of dose; this system has therefore been proposed as a suitable gas dosimeter. The reactants and products are very stable and the results are independent of temperature in the range $-80°C$ to $200°C$ and are also independent of LET. The dosimeter would also be capable of covering a wide range of doses. For doses between 10^4 and 10^7 rad, the dose may be determined from the amount of nitrogen produced. At higher doses, up to 10^9 rad, the dose may be calculated from the amount of nitrogen dioxide formed. This may be determined photometrically without opening the radiation vessel.

At pressures below 50 Torr the mechanism of the radiation-induced decomposition of nitrous oxide is uncertain, and the system is therefore unsuitable for dosimetry. It has been suggested that an ionically initiated chain mechanism operates at these low pressures.

CARBON DIOXIDE

Carbon dioxide is virtually unaffected by ionising radiations. Only in its liquid or solid form are appreciable amounts of carbon monoxide and oxygen detected:

$$CO_2 \longrightarrow\!\!\!\sim\!\!\!\sim\!\!\!\sim\!\!\!\longrightarrow CO + O \qquad (7.36)$$

In the gas phase the very small yields have been attributed to the back reaction:

$$CO + O_3 \longrightarrow CO_2 + O_2 \qquad (7.37)$$

the ozone arising from the reaction of oxygen atoms with oxygen molecules. Other suggestions have been that the carbon monoxide reacts with O_2^+, O_2^- and CO_3 to reform carbon dioxide. At high dose rates there is some net decomposition of gaseous carbon dioxide to carbon monoxide and oxygen, and the stationary concentrations of the products are proportional to the square root of the dose rate. This conclusion is in accordance with the apparent stability of carbon dioxide at low dose rates.

In addition to the formation of carbon monoxide and oxygen, lower polymeric oxides of carbon have been detected on the walls of the vessel. These oxides include C_2O and the carbon suboxide, C_3O_2, which is the anhydride of malonic acid. These polymers appear to catalyse the back reaction to CO_2.

The radiation chemistry of carbon dioxide is of some importance as it is the cooling gas in nuclear power reactors. Here, too, carbon monoxide and the other suboxides are produced, but the substantial back reactions (to CO_2) keep the coolant gas stable.

HYDROCARBONS

Turning now to the hydrocarbon gases, a great deal of work has been done on the radiation chemistry of the aliphatics. These are of great interest to the petroleum industries.

Methane

Methane has been investigated in detail by use of both the mass spectrometer and radical scavengers. The mass spectrometer shows up the ions CH_4^+, CH_3^+, CH_2^+ in the ratio 48 : 40 : 8. In addition, CH_5^+ and $C_2H_5^+$ appear and these must be due to the ion–molecule reactions:

$$CH_4^+ + CH_4 \longrightarrow CH_5^+ + CH_3 \qquad (7.38)$$

$$CH_3^+ + CH_4 \longrightarrow C_2H_5^+ + H_2 \qquad (7.39)$$

More hydrogen is formed by the neutralisation reaction

$$CH_5^+ + e^- \longrightarrow CH_3 + H_2 \qquad (7.40)$$

Radical scavengers such as iodine have shown that the radicals CH_3, H, CH_2 and C_2H_5 are formed in the proportions $70:8·5:8·5:4·5$. The result of all the species interacting and in particular of such radical–radical interaction as

$$2CH_3 \longrightarrow C_2H_6 \qquad (7.41)$$
$$\text{(ethane)}$$

$$2C_2H_5 \longrightarrow C_2H_6 + C_2H_4 \qquad (7.42)$$
$$\text{(ethane) (ethylene)}$$

$$2C_2H_5 \longrightarrow C_4H_{10} \qquad (7.43)$$
$$\text{(butane)}$$

have the effect of producing dimers and even oligomers of the parent molecule. The irradiation of methane, however, leads mainly to hydrogen and ethane, as is shown by the G-values in *Table 7.2*.

Table 7.2 Products of methane irradiation

Product	H_2	C_2H_6	C_3H_8	C_4H_{10}	C_2H_4
G/molecules $(100 \text{ eV})^{-1}$	6	2·1	0·26	0·19	0·13

L

Ethane

In the irradiation of ethane, a similar series of complex reactions among an even greater number of intermediate species form hydrogen, butane, methane, propane and pentane with G-values of 6·8, 1·0, 0·61, 0·54 and 0·54 molecules/100 eV, respectively.

Propane

Irradiation of propane results in hydrogen, methane, acetylene, ethylene, propylene, butanes, pentanes and hexanes. The initial reactions occur as a result of the action of free radicals formed from an ion neutralisation reaction and the dissociation of an excited molecule:

$$C_3H_8^+ + e^- \longrightarrow C_3H_7 + H \qquad (7.44)$$

$$C_3H_8^* \longrightarrow C_3H_7 + H \qquad (7.45)$$

Disproportionation leads to

$$C_3H_7 + C_3H_7 \longrightarrow C_3H_6 + C_3H_8 \qquad (7.46)$$

It is possible to distinguish between the neutralisation and excitation reactions (7.44) and (7.45) by introducing electron scavengers or by carrying out the irradiation in an electric field. Both of these techniques have the effect of suppressing the neutralisation reaction, radical formation occurring only from excited states.

One of the many results of the primary processes is the formation of hydrogen. As will be shown below, a great deal of information can be obtained just by determining this one product quantitatively. Hydrogen may be formed as the result of a hydrogen abstraction reaction involving hydrogen atoms and propane:

$$H + C_3H_8 \longrightarrow H_2 + C_3H_7 \qquad (7.47)$$

In addition, hydrogen may be formed directly as a molecular product:

$$C_3H_8 \longrightarrow H_2 + C_3H_6 \qquad (7.48)$$

Thus, initially, the total yield of hydrogen, $G(H_2)$, is given by

$$G(H_2) = G_{H_2} + G_H \qquad (7.49)$$

It is found experimentally that $G(H_2)$ falls with increasing dose until it eventually reaches a constant value, which is independent of any further increase in dose. This decrease in $G(H_2)$ is attributed to the scavenging of hydrogen atoms by propylene, one of the other products of radiolysis:

$$H + C_3H_6 \longrightarrow C_3H_7 \tag{7.50}$$

This reaction is an addition reaction. As the amount of propylene in the system increases with dose, more and more hydrogen atoms disappear by reaction (7.50) until reaction (7.47) is completely suppressed. At this stage, the total yield of hydrogen must be equal to the molecular yield from reaction (7.48) and

$$G(H_2) = G_{H_2} \tag{7.51}$$

This conclusion is confirmed by initially adding olefins to the system which scavenge the hydrogen atoms and decrease $G(H_2)$ to that constant value achieved at high doses in the absence of scavengers. In the presence of an olefin scavenger, S, some of the hydrogen atoms will react with propane according to reaction (7.47) and some will react with the scavenger according to

$$H + S \longrightarrow SH \tag{7.52}$$

The fraction, f, of the hydrogen atoms reacting with the propane will be given by

$$f = \frac{v_{47}}{v_{47} + v_{52}} \tag{7.53}$$

where v_{47} and v_{52} are the velocities of reactions (7.47) and (7.52), respectively. If one expresses these velocities in terms of rate constants and concentrations of reactants, equation (7.53) becomes

$$f = \frac{k_{47}[H][C_3H_8]}{k_{47}[H][C_3H_8] + k_{52}[H][S]} \tag{7.54}$$

or

$$f = \frac{1}{1 + k_{52}[S]/k_{47}[C_3H_8]} \tag{7.55}$$

As the radiolytic yield of hydrogen atoms is G_H and the fraction of them reacting with propane to produce hydrogen is f, then the radiolytic yield of hydrogen from reaction (7.47) is $G_H f$. The total yield of hydrogen is equal to this quantity plus the yield of molecular hydrogen from reaction (7.48) so that

$$G(H_2) = G_{H_2} + G_H f$$

or

$$G(H_2) = G_{H_2} + G_H \cdot \frac{1}{1 + k_{52}[S]/k_{47}[C_3H_8]} \tag{7.56}$$

Rearranging,

$$G(H_2) - G_{H_2} = G_H \cdot \frac{1}{1 + k_{52}[S]/k_{47}[C_3H_8]} \qquad (7.57)$$

Inverting equation (7.57) gives

$$\frac{1}{G(H_2) - G_{H_2}} = \frac{1}{G_H} \cdot \frac{k_{52}[S]}{k_{47}[C_3H_8]} + \frac{1}{G_H} \qquad (7.58)$$

With a constant concentration of propane, a plot of the left-hand side of equation (7.58) against [S] gives a straight line of intercept $1/G_H$ and slope $k_{52}/G_H k_{47}[C_3H_8]$, from which G_H and k_{52}/k_{47} may be determined. If one of these rate constants is known already, the other may be evaluated. This is a good example of the service which radiation chemistry can give to chemical kinetics in determining rate constants which are difficult or impossible to determine by other means.

Acetylene

Finally, a word may be said about another gaseous hydrocarbon, acetylene, which has been studied mainly because of its high degree of unsaturation. The main product on irradiation appears to be cuprene, a polymer which consists of cyclic hydrocarbons containing aromatic compounds, olefins and small quantities of paraffins (hexane). The other radiation product is benzene, with a G-value of about 5·1 molecules/100 eV. As $G(-C_2H_2)$ is about 70 molecules/100 eV, cuprene appears to be formed by a chain reaction, although the constant yield, independent of dose rate, pressure and temperature, appears to contradict such a process.

Unsaturated hydrocarbons irradiated in the presence of other gases such as hydrogen bromide will form the appropriate addition compound as well as a polymer.

SOLIDS

Investigation into the effects of ionising radiations on solids has led, not only to a greater understanding of radiation chemistry, but also to a better knowledge of solid state physics. In the solid state the ionised and excited species will be much more closely confined than in the gas phase. The distribution of these species in the solid state will thus be very dependent on LET. Furthermore, the atoms and molecules in the neighbourhood of the ionised and excited species are also restricted in the solid state, so that the deposited energy is confined to a very small region in the solid. With radiation of high LET (or where high LET develops, as at the end of a β-particle track), 'hot spots' develop. These hot spots are volumes (about 3 nm in diameter) of intense excitation and ionisation, and the local temperature in such regions can rise to over 1000°C for about 10^{-11} s. The result in the solid is often metastable phases and an acceleration of diffusion processes. Fission products in the uranium metal fuel elements of reactors produce the most intense 'thermal spikes', with temperatures reaching 4000°C in cylinders with radius 10 nm and length 4000 nm (4 μm).

It has been shown in Chapter 3 that all types of ionising radiations give rise to ionisation and excitation through the action of charged particles in inelastic collisions. If the charged particle is a positive ion, it will eventually become neutralised by combination with an electron. This will occur when it has lost most of its original energy and it will then interact with the medium through which it passes by elastic collisions only. No ionisation is produced at this stage, but the elastic collisions with absorber atoms will cause the displacement of such an atom if the absorber is a solid. There is a threshold energy, E_d, for such 'knock-ons' and displacements are observed when the energy of the incident particle is between E_d and E_i, the lowest energy required to produce ionisation in the target. The number of primary displacements, n_p, produced per unit volume of absorber is given by

$$n_p = \phi t n_0 \sigma \qquad (7.59)$$

where ϕ is the flux of incident particles, t is the time, n_0 is the number of target atoms per unit volume and σ is the cross-section for displacement. The total number of displacements will, however, be greater than n_p, as an atom which undergoes a primary displacement may itself cause further displacements. The total number of displacements, N_d, is thus given by

$$N_d = n_p \bar{\nu} \qquad (7.60)$$

where $\bar{\nu}$ is the number of displacements per primary displacement. The effect of the energy of the incident particle on $\bar{\nu}$ is shown in *Figure 7.1*. It can be seen that $\bar{\nu}$ depends on the energy of the incident particle between E_d (c. 25–50 eV) and E_i. Below E_d, $\bar{\nu}$ has the value unity. When the energy of the incident particle is greater than E_d, $\bar{\nu}$ is given by the following relations. For absorbers with mass numbers greater than 45,

$$\bar{\nu} = \frac{\bar{E}}{2E_d} \qquad (7.61)$$

where $\bar{E}$ is the average energy of the incident particle. For absorbers with mass numbers less than 45,

$$\bar{v} = \frac{E_i}{2E_d} \tag{7.62}$$

When the energy of the incident particle is less than E_i, then the main effect which such a particle produces is a primary displacement with a number of subsequent displacements which depends on the

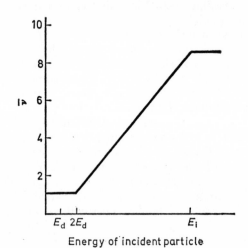

Energy of incident particle

Figure 7.1.   Effect of incident particle energy on displacements

energy which the primarily displaced particle has above E_d. It will be recalled that elastic collisions were discussed in Chapter 3, where the fraction, δ, of the energy of the incident particle which is transferred to the displaced particle is given by equation (3.2) as

$$\delta = \frac{4m_1m_2}{(m_1 + m_2)^2} \cos^2 \theta \tag{3.2}$$

If the energy of the incident particle is E_{in} and the energy transferred to the displaced particle is E_t, then

$$\delta = E_t/E_{in} \tag{7.63}$$

and equation (3.2) may be rewritten as

$$E_t = E_{in} \cdot \frac{4m_1m_2}{(m_1 + m_2)^2} \cos^2 \theta \tag{7.64}$$

It has been pointed out already (Chapter 3) that the maximum amount of energy is transferred when $\theta = 90°$. If we represent this maximum transferred energy as $E_{t(max)}$, then

$$E_{t(max)} = E_{in} \cdot \frac{4m_1m_2}{(m_1 + m_2)^2} \tag{7.65}$$

It will be appreciated that the minimum amount of energy which may be transferred will be zero for a collision which is just a miss when $\theta = 0°$. The average energy transferred $\bar{E}_t$ is thus given by

$$\bar{E}_t = \tfrac{1}{2}E_{t(max)} = E_{in} \cdot \frac{2m_1m_2}{(m_1 + m_2)^2} \tag{7.66}$$

If, for example, 2 MeV neutrons impinge on hydrogenous material, the value of the average energy transferred to a primarily displaced proton would be given by

$$\bar{E}_t = 2 \text{ MeV} \times \frac{2 \times 1 \times 1}{(1 + 1)^2} = 1 \text{ MeV}$$

If the target atoms were carbon, copper or uranium the displaced atoms would have energies of 280 keV, 61 keV and 17 keV, respectively. The ionisation energy of an atom in keV is approximately equal to its mass number, so the values of E_i for copper and uranium are approximately 65 keV and 238 keV, respectively. In view of these figures and those given above, it will be appreciated that when copper or uranium is bombarded with 2 MeV neutrons, the average energy transferred is well below the ionisation energies of these metals and only displacement can be expected as a result of interaction.

The total number of displacements is dependent on the quantities E_i and $\bar{E}_t$, and it will have been observed from equation (7.66) that this latter quantity is dependent upon the mass of the incident particle. Thus both neutrons and γ-rays can cause displacements, but in the latter case the displacements are produced by the secondary electrons which are released. Owing to the difference in the masses of a neutron and an electron, neutrons cause many more displacements than do γ-rays under comparable conditions. If copper is bombarded for 1 month with 2 MeV neutrons at a flux of 10^{13} neutrons cm^{-2} s^{-1}, then 10^{-1} displacements per copper atom are observed. If 1·3 MeV γ-rays are used for the same period at the same flux, only 10^{-6} displacements per copper atom occur.

From all the above it has been seen that displacements occur only if the energy which the target atom receives is greater than E_d. If, after collision, the energy of the incident particle is less than E_d, it will remain in the target material. The over-all situation can be represented by the model of Kitchen and Pease shown in *Figure 7.2*. It will therefore be seen that the primary processes which take place in solids as a result of the absorption of ionising radiations consist of the usual ionisations and excitations observed in other phases,

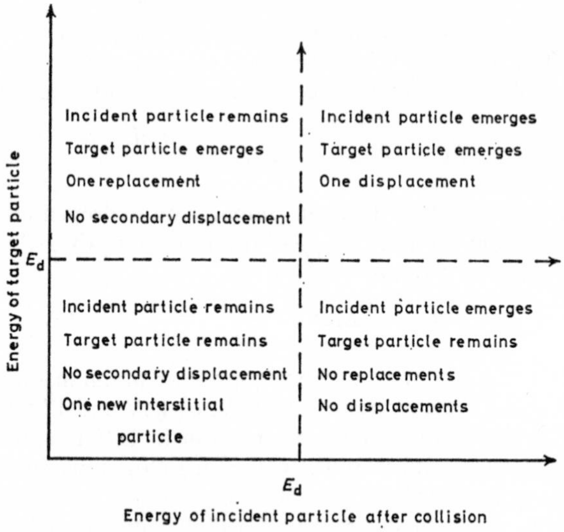
Incident particle remains
Target particle emerges
One replacement
No secondary displacement
Incident particle emerges
Target particle emerges
One displacement
Incident particle remains
Target particle remains
No secondary displacement
One new interstitial particle
Incident particle emerges
Target particle remains
No replacements
No displacements
Energy of target particle
E_d
E_d
Energy of incident particle after collision

Figure 7.2. Kitchen and Pease model

together with the additional displacement and the formation of interstitial species. The latter two phenomena manifest themselves in a variety of ways.

FORMATION OF COLOUR CENTRES

In alkali halide crystals where alternate ions carry opposite charges, displacement effects can be more clearly defined than elsewhere. The result of these displacements in alkali halide crystals is the formation of colour centres, which affect the absorption of light. *Figure 7.3* shows the variation in absorbance with wavelength due to a variety of defects arising as a result of displacements.

The increase in absorbance at certain wavelengths can be ascribed to the formation of particular defects. The displacement of a negative ion from the lattice leaves a negative ion vacancy. Irradiation also causes electrons to be ejected from the ions in the crystal and sometimes it happens that such an electron becomes trapped in a negative ion vacancy. This gives rise to an F-centre. The trapping of two electrons constitutes an F'-centre and the creation of two negative ion vacancies with the subsequent trapping of one or two electrons gives rise to an R-centre and an R_2-centre, respectively. These situations are represented in *Figure 7.4*.

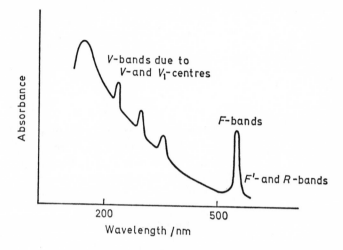

Figure 7.3. Absorbance due to defects (KCl)

The displacement of a positive ion leaves a positive ion vacancy. The loss of an electron from an adjacent negative ion converts the ion into a radical. The combined positive ion vacancy and radical constitutes a V_1-centre and the radical may be regarded as a bound positive hole. The formation of a V_1-centre is illustrated in *Figure 7.5*. In the V_1-centre the loss of the electron by the negative ion is shared by the five halide ions surrounding the positive ion vacancy.

Other solids show colour changes as a result of exposure to radiation. The darkening of silver-activated phosphate and cobalt borosilicate glasses has already been mentioned (Chapter 5), as were the changes in various coloured plastics. Glasses, although solid, have a liquid-like structure and the electrons which are released by the ionising radiation will tend to reduce any cations present. Thus, silver ions are converted to silver, manganic ions to manganous ions, ferric ions to ferrous ions and ceric ions to cerous ions in glasses. Free electrons are also believed to contribute to the darkening of glasses and their presence can be measured by e.p.r.

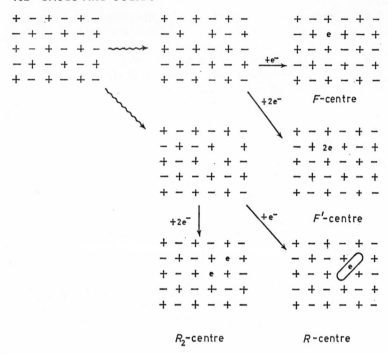

Figure 7.4. Formation of F- and R-centres

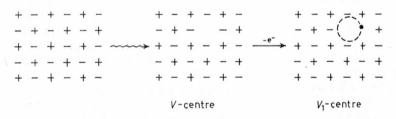

Figure 7.5. Formation of a V_1-centre

It is not certain whether atomic displacements in glasses contribute to these discolorations. The colour centres are subject to ageing and removal through heating. It is assumed that, under these circumstances, the metal ions are oxidised back to their original state and the electrons return to their former sites.

Such effects are also observed with frozen solutions irradiated at very low temperatures. Hydrogen atoms can be detected by e.p.r. in irradiated acid and alkaline ice. Free radical behaviour in irradiated organic glasses has been investigated below −77°C. As the temperature is raised, the radicals trapped in these glasses are converted into ionic or covalent species as a result of electron migration and pairing. All these reactions can be followed by e.p.r. and spectroscopy. Besides e.p.r. and the optical changes, two other physical phenomena can be affected when solids are subjected to ionising radiation. One is electrical conductivity and the other is the dimensions of the solid.

ELECTRICAL CONDUCTANCE

In the solid state, the sharply defined energy levels of a single atom are broadened into bands of closely packed energy levels due to the influence of neighbouring atoms. Electrons in the lower filled bands cannot move under the influence of an electric field, because to be accelerated by the field they would have to move into a higher energy level. If an electron is in a lower level, the levels immediately above will be completely filled and unable to accommodate extra electrons. Electrons in an only partially filled upper band can move to a higher level and move under the influence of an electric field. Moreover, if an empty band overlaps a filled band, electrons may be promoted to the empty band and then be free to move in an applied field. Thus, electrical conductors are characterised by partial filling of bands or by overlap of the topmost bands. The bands containing the fixed electrons are, of course, the *valence bands*, and those accommodating electrons moving under the influence of an electric field are called *conduction bands*. In a non-conductor there is a large energy gap between the filled valence band and the conduction band, so there is little possibility of the promotion of an electron to the conduction band. This situation is illustrated in *Figure 7.6*.

In semiconductors there are two situations to be considered. First, there are *intrinsic* semiconductors, in which the valence band is filled but the conduction band is separated from it by only a small energy gap, across which electrons can be promoted. This process leaves positive holes in the valence band and both the holes and electrons will be mobile under the influence of an electric field. There are also *impurity* semiconductors, in which the energy gap is large but the incorporation of an impurity either introduces vacant levels (acceptor levels) just above the valence band or introduces filled levels (donor levels) just below the conduction band. These

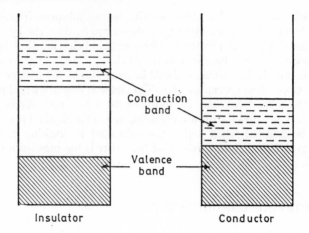

Figure 7.6. Valence and conduction bands in insulators and conductors

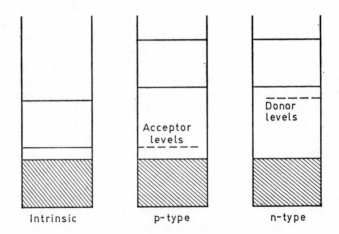

Figure 7.7. Semiconductor energy levels

types of impurity semiconductors are known as *p-type* and *n-type*, respectively. *Figure 7.7* illustrates the position.

Considering now the effects of radiation, metals irradiated at normal temperatures will be little affected. Any ionisation will be quickly counteracted by the floating cloud of electrons in the conduction band. At very low temperatures ($-193°C$), the displacement of metal atoms does cause a slight increase in resistivity. At all temperatures, displacements and the production of interstitial atoms increase the brittleness of a metal and decrease its ductility.

Much greater changes in electrical conductivity are observed in semiconductors and non-conductors. In the intrinsic semiconductor, few electrons are free to roam across the atoms. Displacements may give rise to additional trapping centres, which may reduce the small amount of current the semiconductor can carry. Opposed to and swamping this effect is the concurrent ionisation and excitation which causes the release of large quantities of electrons and the formation of positive holes. As the latter radiation effects are predominant and directly related to the energy of the incident particle or ray, this system is, in fact, used for the detection and measurement of radiation in the form of p–n junctions and surface barrier detectors. With these devices, a portion of the detector is drained completely of any floating electrons and mobile positive holes. This converts it essentially into a non-conductor. Any ionisation taking place in this 'depleted zone' is wholly due to the absorption of radiation and can be accurately measured. Such semiconductor devices are the solid equivalent of the ionisation chamber.

DIMENSIONAL CHANGES

For dimensional changes to be observed in solids, the radiation dose has to be quite large and the irradiating particles have to be of substantial mass. Such conditions exist in nuclear reactors, and dimensional changes were observed in the uranium nuclear fuel. In this case, the bombarding particles were neutrons and fission fragments. Metallic uranium exists in three allotropic forms, α, β and γ, the transition temperatures for $\alpha \rightarrow \beta$ and $\beta \rightarrow \gamma$ being 660 and 760°C, respectively. In the α-phase, heating and irradiation cause distortion, because the crystal expands in two of its crystal planes and contracts in the third. With nuclear fuel this effect can be minimised by orienting the metal crystals in every direction prior to irradiation. In practice, it is achieved by avoiding any severe work to be done on the cold metal, as this orientates the crystals. Alternatively, if the metal has to be worked, this is done at high temperatures. As a further precaution the uranium metal in the reactor is constricted by canning it in 'Magnox', an aluminium–magnesium alloy. This is also necessary to prevent the escape of fission products.

Another nuclear power reactor component which shows a large dimensional change is the graphite moderator. This change occurs when the graphite is irradiated at low temperatures ($c.$ 30°C). In the graphite structure, shown in *Figure 7.8*, the carbon atoms in the rings are separated by a distance of 0·142 nm and the ring layers are 0·355 nm apart. The introduction of interstitial carbons due to displacement reactions from neutrons can cause a separation between the planes of 0·6 nm and gives a total dimensional change of 6%. As the radiation discoloration in glass can be removed by heating,

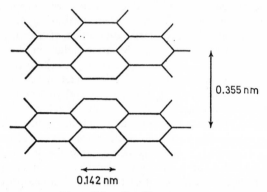

0.355 nm

0.142 nm

Figure 7.8.   Graphite structure

so here, in the same way, the displaced carbon atoms can be returned to their original position. As this process occurs, however, the 'stored energy' due to the displacements is given out in the form of heat and a sudden increase in temperature is observed. This storage of energy by displacement, amounting to about 42 J g^{-1} for 10^{20} neutrons cm^{-2} s^{-1}, is known as the *Wigner energy* and the effect of its release is shown in *Figure 7.9*.

Early nuclear reactors were run at low temperatures purely for the production of plutonium. From time to time, it was necessary to rectify the distortion in their moderator. Modern nuclear reactors are used mainly to produce electric power and are run at as high a temperature as possible. Thus almost as soon as a carbon atom is displaced, it returns to its original position or, if it cannot do that, at least no Wigner energy is stored.

CHEMICAL CHANGES

In addition to the physical changes described above many solids show some chemical change. These, however, are often difficult to detect; because if the solid is dissolved or melted for analytical purposes, reverse or secondary reactions may take place, or both.

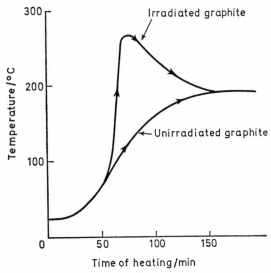

Figure 7.9. *Wigner energy release*

The formation of nitrite ion and oxygen gas from the irradiation of alkali metal nitrates had been discovered very early (1907). It was found later that $G(NO_2^-)$ and $G(O_2)$ varied with the temperature of the irradiated potassium nitrate. The reactions occurring are probably

$$NO_3^- \longrightarrow\!\!\!\sim\!\!\sim\!\!\longrightarrow (NO_3^-)^* \qquad (7.67)$$

$$(NO_3^-)^* \longrightarrow NO_2^- + O \qquad (7.68)$$

$$O + NO_3^- \longrightarrow NO_2^- + O_2 \qquad (7.69)$$

The occluded oxygen gives the crystal a milky appearance. Yields of nitrite ions are dependent on the size and charge of the cation and the degree of hydration.

Dry azides decompose when subjected to ionising radiation to give nitrogen. The over-all process in sodium azide can be written

$$NaN_3 \longrightarrow\!\!\!\sim\!\!\sim\!\!\longrightarrow \tfrac{3}{2} N_2 + Na \qquad (7.70)$$

and is said to occur through the formation of positive holes, an azide radical, N_3, constituting a positive hole:

$$N_3^- \longrightarrow\!\!\!\sim\!\!\sim\!\!\longrightarrow N_3 + e^- \qquad (7.71)$$

$$N_3 + N_3^- \rightarrow 3N_2 + e^- \qquad (7.72)$$

Another anion which decomposes on irradiation is perchlorate. In this case also the reaction is temperature dependent. The main products are oxygen, chloride ions and chlorate ions formed probably through the ClO_4 and ClO_3 radicals as intermediates. Irradiation of potassium chlorate gives chlorides, chlorites and oxygen.

In a number of salts, exposure to ionising radiation results in a reduction of the metal cations to the metallic form. The most important of these processes is the reduction of the silver ion in silver bromide. This is reduced to form silver in wrapped photographic films by penetrating radiation. Another reduction process occurs in the irradiation of diamminodichloro platinum $[Pt(NH_3)_2Cl_2]$ and this results in the formation of platinum metal.

A number of chemical reactions with solids have been studied for their irradiation 'after-effects'. These are known as *topochemical reactions* and are the surface-chemical equivalent of the light emission in radiation-induced thermoluminescence in lithium fluoride and calcium fluoride. Silver oxalate is an example of a compound whose thermal decomposition is strongly affected by prior exposure to ionising radiation. Pre-irradiation can cause the silver oxalate to explode on heating. This is thought to be due to the release of electrons (in the solid) which are captured at sites vacated by anions. Such spots would then act as multi-initiating centres for multichain thermal decomposition.

The effect of ionising radiation on heterogeneous catalysts is not clear, because in some processes it increases the reaction rate (e.g. conversion of *ortho*-H_2 to *para*-H_2; $H_2 + D_2 \rightarrow 2HD$), but in others the rate is decreased (e.g. oxidation of SO_2 with V_2O_5; hydrogenation of ethylene on ZnO). In a few cases where heterogeneous systems have been irradiated, it was found that some of the energy absorbed in the solid was transferred to the monolayer or the whole of the other phase.

A great deal of work has been carried out at low temperatures (from 4·2 K up to room temperature), e.p.r. being used to study the behaviour of transient radicals. Also, the reactive decay, and conversion of the radicals to other radicals in the solid (frozen) state, can be followed if a gradual 'warm-up' from near absolute zero to room temperature is allowed. More than 120 compounds have been investigated by this technique. These have included hydrocarbons, alcohols, organometallic compounds, organic acids, amides, lactams, organic chlorides, amino acids and polymers.

8
Organic systems

When ionising radiations are absorbed by organic materials, the primary effects are similar to those occurring in water. Ionisation and excitation will take place, resulting in the formation of positive ions, electrons and radicals. These processes were discussed when the irradiation of hydrocarbon gases (e.g. CH_4, C_2H_6) was considered in Chapter 7. In those cases the production of the primary species led to radical–radical, ion–molecule and neutralisation interactions. Because they were in the gas phase, the transient species did not experience a cage effect such as would be expected in similarly constituted liquids (e.g. pentane and hexane). In the liquid phase the ejected electron is caged with its parent ion, so recombination neutralisation is likely. In addition, electron interaction with the surrounding molecules is less likely in organic systems because their dielectric constants are smaller than that of water. In the event of a neutralisation reaction between the electron and its parent ion, the recombined molecule will be in a highly excited state—much more excited than a molecule which has obtained its excitation energy directly from the radiation beam. The highly excited molecule will, of course, decompose into radical or molecular products:

$$A \xrightarrow{\hspace{2cm}} A^+ + e^- \tag{8.1}$$

$$A^+ + e^- \longrightarrow A^{**} \tag{8.2}$$

$$A^{**} \longrightarrow \text{radical or molecular products} \tag{8.3}$$

M

The less highly excited molecules which are formed directly may decompose or return to the ground state:

$$A \longrightarrow\!\!\!\sim\!\!\!\sim\!\!\!\sim\!\!\!\longrightarrow A^* \tag{8.4}$$

$$A^* \longrightarrow \text{radical or molecular products} \tag{8.5}$$

$$A^* \longrightarrow A \tag{8.6}$$

It can be expected that there will be a difference in the energy states of the radicals formed by reaction (8.3) and reaction (8.5). Radicals from the highly excited neutralised molecules will naturally be in a more highly excited state than the radicals coming from the radiation-excited molecule.

This is, then, the real difference between irradiating an organic compound directly or in aqueous solution. In the latter case the organic compound is the solute and subject to attack by hydroxyl radicals, hydrogen atoms and solvated electrons. There may also be interaction with the molecular products, hydrogen and hydrogen peroxide. No direct ionisation of the organic solute occurs, and organic radicals are formed solely as a result of dehydrogenation and addition reactions. Also, excited organic states are not formed by absorption of the radiation (as in photochemistry). On the other hand, direct irradiation of an organic compound, as discussed above, produces excited states and highly excited states, organic ions and electrons. *Table 8.1* shows some of the general differences which can be expected in the irradiation of polar and non-polar compounds.

Table 8.1 Reactions in polar and non-polar media

Species and time scale	Polar media (aqueous)	Non-polar media (hydrocarbon)
Intermediate species, $10^{-18} - 10^{-12}$ s	excited states of water, ions, electrons radicals	excited states of organic compounds, ions, electrons, radicals
Interaction of ions and electrons, $10^{-12} - 10^{-6}$ s	interaction with media molecules leading to the formation of radicals	little interaction with media molecules; instead, neutralisation recombination in cage leading to highly excited radicals
Interaction of radicals, 10^{-6} s onwards	little interaction with media molecules instead, some caging and self-interaction and interaction with solutes	interaction with media molecules, LET effect usually unimportant

Radiation products formed directly or as a result of radical inter-action can be distinguished by means of the addition of radical scavengers. These will, of course, reduce the amount of products which are formed by the action of free radicals.

HYDROCARBONS

When liquid hydrocarbons are irradiated in the absence of oxygen, the main product is usually hydrogen gas. The ionisation and excita-tion reactions may be represented

$$RH_2 \longrightarrow\!\!\!\sim\!\!\!\sim\!\!\!\sim\longrightarrow RH_2^+ + e^- \qquad (8.7)$$

$$RH_2 \longrightarrow\!\!\!\sim\!\!\!\sim\!\!\!\sim\longrightarrow RH_2^* \qquad (8.8)$$

Following these initial processes there can be several radical reactions, including addition, disproportionation and combination reactions:

$$RH_2^* \longrightarrow RH + H \qquad (8.9)$$

$$RH_2^* \longrightarrow R + H_2 \qquad (8.10)$$

where R is an unsaturated product.

$$H + RH_2 \longrightarrow RH + H_2 \qquad (8.11)$$

$$H + RH_2 \longrightarrow RH_3 \qquad (8.12)$$

$$RH + RH \longrightarrow R_2H_2 \qquad (8.13)$$

$$RH + RH \longrightarrow RH_2 + R \qquad (8.14)$$

$$H + H \longrightarrow H_2 \qquad (8.15)$$

$$H + RH \longrightarrow RH_2 \qquad (8.16)$$

$$H + RH \longrightarrow H_2 + R \qquad (8.17)$$

The ions can also take part in neutralisation and ion–molecule reactions:

$$RH_2^+ + e^- \longrightarrow RH_2^{**} \qquad (8.18)$$

In this case the product subsequently decays to RH_2^* or decomposes.

$$RH_2^+ + RH_2 \longrightarrow R_2H_2^+ + H_2 \qquad (8.19)$$

In the mass spectra of hydrocarbons, a large number of ions are observed. In the case of hexane, for example, these vary from CH_3^+ to $C_5H_{11}^+$ Any of these ions can form the primary ion RH_2^+ or become neutralised, giving highly excited species as described above.

Alongside the hydrogen gas, then, there are produced molecular fragments of the substance irradiated as well as unsaturated and higher relative molecular mass compounds. In fact, so many are the radiation products from compounds of high relative molecular mass, that only in a few cases have they all been identified and their yields measured. Thus the mechanisms leading to even the main products can never be stated completely. This is true for many organic systems besides pure hydrocarbons. The over-all effect of energy brought into an organic system by ionising radiation is to break covalent bonds somewhat indiscriminately. Of course, where there is a great difference in bond strength then, naturally, the weaker bond will break first, as, for example, the carbon–halogen bond in organic halides. The bond energies of C—H and C—C, however, are so near to each other (410·5 and 334·4 kJ mol^{-1}, respectively) that they will be broken roughly in the ratio of their abundance.

The radiation yields of hydrogen from straight chain hydrocarbons up to hexadecane lie in the region of $G \approx 5$ molecules/100 eV. With configurations other than straight chain, when not all the carbon atoms carry their maximum quota of hydrogen atoms, the yield of hydrogen is decreased. In 2,2-dimethyl butane, for example, there are fewer C—H bonds than in hexane but there are more C—CH$_3$ bonds. As a result of this difference, the radiolysis of 2,2–dimethyl butane gives a higher yield of methane, one of the fragmentary products, while the yield of hydrogen falls to 2 molecules/100 eV. This point is illustrated in *Table 8.2*.

Table 8.2 Variation of yields with structure

Compound	$G(H_2)$ *molecules/100* eV	$G(CH_4)$ *molecules/100* eV
Hexane	5·0	0·15
2,2-Dimethyl butane	2·0	1·2

Hexane itself shows at least 16 radiation products. Besides the hydrogen and methane mentioned above, the other two main products are the dimer, $C_{12}H_{26}$ ($G = 2 \cdot 0$ molecules/100 eV), and the *trans*-unsaturated compound, C_6H_{12} ($G = 1 \cdot 2$ molecules/100 eV). When hydrocarbon chains become really long, as in dodecane, and the radiation dose becomes very high (*c.* $6 \cdot 8 \times 10^8$ rad), sufficient cross-linking dimerisations occur to cause gel formation.

Irradiation of cyclic hydrocarbons also produces hydrogen gas (e.g. $G(H_2) = 5 \cdot 5$ molecules/100 eV from cyclohexane), but other products are fewer than in the straight chain analogues. This smaller number is probably due to the fact that all the C—H bonds in the cyclic compound are the same, which reduces the variation in the type of reactions following bond breakage. As before, the dimer and the unsaturated analogue of the parent compound are among the main products. In the case of cyclohexane these would be bicyclo-hexyl ($G = 1.95$ molecules/100 eV) and cyclohexene ($G = 3.27$ molecules/100 eV). The mechanisms leading to the formation of hydrogen, dimer and unsaturated compound are similar to those described for the straight chain hydrocarbons.

In the presence of oxygen, all radical intermediates will interact with oxygen to form peroxy radicals, thus reducing the yields of dimers and unsaturateds to negligible amounts. Variation of reaction products with temperature and pressure suggest that organic hydro-peroxides and peroxides are among the intermediates. The types of reaction expected in the radiolysis of an organic molecule, RH_2, in the presence of oxygen are as follows:

$$RH_2 \longrightarrow\!\!\!\sim\!\!\!\sim\!\!\!\sim\longrightarrow RH + H \qquad (8.20)$$

$$RH + O_2 \longrightarrow HRO_2 \qquad (8.21)$$

$$H + O_2 \longrightarrow HO_2 \qquad (8.22)$$

$$HO_2 + HO_2 \longrightarrow H_2O_2 + O_2 \qquad (8.23)$$

$$HRO_2 + RH \longrightarrow HRO_2RH \qquad (8.24)$$
$$\text{organic peroxide}$$

$$HRO_2 + HRO_2 \longrightarrow HRO_2RH + O_2 \qquad (8.25)$$

$$H + HRO_2 \longrightarrow HRO_2H \qquad (8.26)$$
$$\text{organic hydroperoxide}$$

$$H + RH_2 \longrightarrow H_2 + RH \qquad (8.27)$$

$$HRO_2 + RH_2 \longrightarrow HRO_2H + RH \qquad (8.28)$$

Carbon monoxide, carbon dioxide and water are formed on the irradiation of hexane in the presence of oxygen, while at higher temperatures epoxides, 2,5-dimethyltetrahydrofurane, acetone and methanol appear. In the case of cyclohexane, irradiation in the presence of oxygen yields the products cyclohexanol, cyclohexanone, cyclohexylhydroperoxide and cyclohexylperoxide.

In comparison with the saturated cyclic hydrocarbons, benzene is much more radiation-resistant. This is probably due to the π-bond orbitals, which can spread the excitation energy around the benzene ring and so give the molecule an opportunity to dissipate the energy before bond breakage occurs. Thus the hydrogen yield from benzene is much lower [$G(H_2) = 0.036$ molecules/100 eV] than that from the corresponding saturated hydrocarbon and the methane yield is nonexistent. Even when an aliphatic chain or group is attached to the ring, little hydrogen is produced. The ring acts as a protective agent for the side chain. The only products of any consequence which arise from the irradiation of benzene are a mixed polymer ($G = 0.75$ molecules/100 eV) and acetylene ($G = 0.02$ molecules/100 eV). The radiation protective effect which benzene has on any substance irradiated in its presence is due to 'riding' the absorbed energy round its π-bond orbital. In this, benzene differs from other protective agents (e.g. cystamine), which compete for and interact with free radicals before these can do any damage. A third form of radiation protection is based on the removal of oxygen from the system. This prevents the formation of organic peroxides, thus reducing the amount of oxidative degradation.

In addition to its radioprotective action, the aromatic ring has another property of interest to radiation and radiochemists. If an aromatic compound is used as the solvent in a solution, the radiation energy which the aromatic compound is capable of holding (for longer than its aliphatic counterpart) can be passed from molecule to molecule until it is finally passed to a solute molecule. If the solute molecule is a scintillating phosphor (e.g. p-terphenyl or 2,5-diphenyloxazole) in a solvent of toluene, a proportion of the energy absorbed will be given out as visible light. The light flashes (scintillations) can be picked up and counted with a photomultiplier tube. These liquid scintillator solutions can be used as detectors for measuring radioactivity. They are particularly suitable for measuring weak beta emitters; the whole of the radiation energy is held in solution. Alternatively, solid scintillators can be made in which polystyrene takes the place of toluene. These can be machined into any suitable shape and are thus very versatile.

One interesting feature of benzene irradiation is that the yields of the products, acetylene and hydrogen show an LET effect. These substances are thought to arise from the interactions of excited species:

$$C_6H_6^* + C_6H_6^* \longrightarrow C_{12}H_{10} + H_2 \qquad (8.29)$$

$$C_6H_6^* + C_6H_6^* \longrightarrow nC_2H_2 + \text{radical or}$$
$$\text{stable products} \qquad (8.30)$$

The recombinations are more likely to occur at high LET, when the local concentration of the excited intermediates is high. At low LET, deactivation by collision will be the more likely process:

$$C_6H_6^* + C_6H_6 \longrightarrow 2C_6H_6 \qquad (8.31)$$

Irradiation of aliphatic unsaturated substances (for example, vinyl compounds such as methyl methacrylate and substituted vinyl compounds such as styrene) leads to the opening of the double bond, with the resultant formation of addition compounds. When only hydrogen and hydrocarbons are present, the bonds become saturated. In the presence of other substances (for example, halogens or oxygen) they form the corresponding addition compounds. Occasionally *cis–trans* isomerisation is observed. Compared with their fully saturated counterparts, aliphatic unsaturated compounds yield on irradiation less hydrogen and more compounds of high relative molecular mass. All these findings are in keeping with the results obtained from conventional organic chemistry.

POLYMERISATION

Under the right conditions, long-chain addition polymers can be formed by the normal addition polymerisation propagation steps. The radiation acts simply as an initiator producing free radicals (usually hydrogen atoms) and ions in the system. The *initiating step* in free radical polymerisation is the production of free radicals by the radiation and it may be represented

$$A \longrightarrow 2R^{\cdot} \tag{8.32}$$

where A is any molecule present in the system and may be either the monomer itself or a solvent. $R^{\cdot}$ represents a primary radical which can then go on to react with an unsaturated monomer molecule to form a larger radical. If the primary radical were a hydrogen atom, this step could be written

$$H + {>}C{=}C{<} \longrightarrow {>}CH{-}\dot{C}{<} \tag{8.33}$$

For convenience, equation (8.33) may be written in the more general form

$$R^{\cdot} + M \longrightarrow RM^{\cdot} \tag{8.34}$$

where M is a monomer molecule. The radical RM* will continue to grow by reacting with further monomer molecules:

$$RM^{\cdot} + M \longrightarrow RM_2^{\cdot} \tag{8.35}$$

This reaction is part of the *propagation* process, which can be written in the general form

$$RM_n^{\cdot} + M \longrightarrow RM_{n+1}^{\cdot} \tag{8.36}$$

Radicals disappear either by combination,

$$RM_n^{\cdot} + RM_m^{\cdot} \longrightarrow P_{n+m} \qquad (8.37)$$

or by disproportionation,

$$RM_n^{\cdot} + RM_m^{\cdot} \longrightarrow P_n + P_m \qquad (8.38)$$

where P_{n+m}, P_n and P_m represent dead polymer molecules.

If the polymer chains are long, the consumption of monomer is virtually limited to the propagation step (reaction 8.36), and the rate of polymerisation is given by

$$\text{Rate} = -d[M]/dt = k_p[RM_n^{\cdot}] [M] \qquad (8.39)$$

where k_p is the rate constant of the propagation step. Under stationary state conditions, the rate of formation of radicals will be equal to their rate of disappearance. Suppose that the rate of formation by the initiating reaction (8.32) is v_i. Radicals disappear by reactions (8.37) and (8.38), and the velocity of each of these reactions may be written $k_t[RM_n^{\cdot}]^2$, where k_t is the rate constant of these *termination steps*. Thus

$$v_i = k_t [RM^n]^2 \qquad (8.40)$$

or

$$[RM^{\cdot}] = (v_i/k_t)^{\frac{1}{2}} \qquad (8.41)$$

Substituting from equation (8.41) into equation (8.39),

$$\text{Rate} = k_p \left(\frac{v_i}{k_t}\right)^{\frac{1}{2}} [M] \qquad (8.42)$$

The rate of formation of radicals, v_i, will, of course, be proportional to the rate at which energy is absorbed by the system, i.e. the dose rate. It will be seen therefore from equation (8.42) that

$$\text{Rate} \propto (\text{dose rate})^{\frac{1}{2}} \qquad (8.43)$$

Deviations from this relationship can occur under certain conditions. The general kinetic scheme described above is governed by the competition for primary radicals. These radicals can react in three possible ways. First, they can initiate polymerisation as shown above in reaction (8.34). Second, they could react with a polymeric radical in a termination step; and third, the primary radicals could recombine with themselves. The three competing reactions are thus

$$R^{\cdot} + M \longrightarrow RM^{\cdot} \qquad (8.34)$$

$$R^{\cdot} + RM_n^{\cdot} \longrightarrow P_n \qquad (8.44)$$

$$R^{\cdot} + R^{\cdot} \longrightarrow R_2 \qquad (8.45)$$

If the concentration of monomer is high with respect to the radical concentration, all the primary radicals will be consumed by reaction (8.34) and equation (8.43) will be valid. With higher dose rates, however, the concentration of radicals will increase, and reactions (8.44) and (8.45) will thus become important. These reactions have not been considered in the derivation of equation (8.43) and it is not surprising, therefore, that deviations from this relation occur at high dose rates.

Deviations from equation (8.43) also occur when the system gels or the polymer comes out of solution. Under these conditions, the mobility of the polymeric radicals is restricted and the termination steps (8.37) and (8.38) are retarded. As a result, the concentration of radicals increases and the rate of polymerisation is accelerated. All radical scavengers (chain inhibitors) will interfere with the process, the most notable one being oxygen.

The products resulting from polymerisation initiated by radiation differ slightly from those initiated by other means but these differences may be of industrial importance. Thus polymerisation initiated by radiation may give rise to a product with higher distortion temperatures and greater density.

It may be more efficient to dissolve the monomer in a solvent before irradiation, as the radiolytic yield of radicals from the solvent may be greater than that of the pure monomer. For example, styrene is more efficiently polymerised by irradiation in aqueous emulsion because the radiolytic yield of radicals from water is greater than that of styrene. The presence of the aromatic ring in the styrene molecule exerts some protective effect.

There are instances when penetrating radiations are the only means of initiating a polymerisation process. The first is under conditions of low temperature (i.e. down to $-78°C$) with, for example, isobutylene. This is indicative, however, of an ionic rather than a free radical mechanism. Where either free radical or ionic (or both) processes contribute to polymerisation, their relative importance can be guaged by studying the rate of monomer conversion at various temperatures. Free radical polymerisation is retarded at low temperatures. The other condition where initiation by ionising rays scores over conventional means is in solid state polymerisations. Here the monomer is in the solid form either naturally or as a result of freezing. In the latter case it is not certain whether the polymerisation of the solid actually occurs at the time of irradiation along the tracks, or later, on warming to room temperature. When the initiating radiation dose is high, some of the cross-linking or degradation effects observed when polymers are irradiated, will be found in the final polymer (see next section).

When copolymers are to be formed, mixtures of the individual monomers are irradiated.

IRRADIATION OF SOLID POLYMERS

Irradiation of solid polymers causes two main chemical effects. In some linear polymers, irradiation leads to *cross-linking* of the linear chains so that the relative molecular mass increases steadily with dose. This process may be represented

$$(8.46)$$

Cross-linking probably occurs through the removal of a hydrogen atom from one of the chains by irradiation. This leaves a radical site on the chain and the hydrogen atom produced will probably abstract another hydrogen atom from an adjacent chain. The two radical sites left on the adjacent chains can then recombine to form a cross-link:

$$\sim\!\!\sim\!CH_2\!\sim\!\!\sim$$

$$\sim\!\!\sim\!CH_2\!\sim\!\!\sim$$

$$(8.47)$$

$$+ H$$

$$\sim\!\!\sim CH\!\sim\!\!\sim$$
$$\bullet$$
$$\sim\!\!\sim CH_2\!\sim\!\!\sim$$

$$(8.48)$$

$$+ H_2$$

$$\sim\!\!\sim CH\!\sim\!\!\sim$$
$$\bullet$$
$$\sim\!\!\sim CH\!\sim\!\!\sim$$

$$(8.49)$$

$$\sim\!\!\sim CH\!\sim\!\!\sim$$
$$|$$
$$\sim\!\!\sim CH\!\sim\!\!\sim$$

The other chemical effect produced by irradiation is *degradation* of the polymer chains. Here the long polymer chains undergo scission and the process may be written

$$\text{(8.50)}$$

Several mechanisms have been adduced to account for chain scission, but none is entirely satisfactory and as yet the mechanism is not well understood.

It has been found in practice that polymers containing the structures $\{CH_2 \text{—} CH_2\}$ or $\{CH_2 \text{—} CHR\}$ mainly undergo cross-linking reactions, while those containing the structures $\{CH_2 \text{—} CR_1R_2\}$ tend to undergo degradation.

In all the polymers irradiated so far, there has always been some gas evolution observed and an analysis of these gaseous products is sometimes useful in giving an understanding of the mechanisms involved. In many polymers the amount of unsaturation increases on irradiation; but in some rubbers which originally contain a large number of double bonds, the unsaturation decreases on irradiation.

Irradiation in the presence of oxygen leads, in general, to the degradation of polymers, even if they have structures which predominantly cross-link in the absence of oxygen. In those polymers which normally undergo degradation, the presence of oxygen leads to enhanced degradation. One suggestion for the mechanism of such a process may be written

$$\text{—CH}_2\text{—CH}_2\text{—} \longrightarrow \text{—CH}_2\text{—}\overset{\bullet}{\text{C}}\text{H —} + \text{H} \qquad \text{(8.51)}$$

$$\text{—CH}_2\text{—}\overset{\bullet}{\text{C}}\text{H —} + \text{O}_2 \longrightarrow \text{—CH}_2\text{—CH—} \qquad \text{(8.52)}$$
$$\underset{\overset{|}{\text{O}_2^{\bullet}}}{}$$

$$\text{—CH}_2\text{—CH—} \longrightarrow \text{—C}\overset{\displaystyle O}{\underset{\displaystyle H}{\diagup}} + \overset{\bullet}{\text{O}}\text{CH}_2\text{—} \qquad \text{(8.53)}$$
$$\underset{\overset{|}{\text{O}_2^{\bullet}}}{}$$

It is common to be able to detect carbonyl and carboxyl groups in polymers which have been irradiated in the presence of oxygen.

It will be appreciated that, in addition to degradation, peroxides and hydroperoxides can be formed by reactions such as

$$\begin{array}{ccc} \sim\!\!\sim\!\!CH_2\!-\!\underset{\underset{\bullet}{O_2^{\bullet}}}{CH}\!\!\sim\!\!\sim & & \sim\!\!\sim\!\!CH_2\!-\!\underset{O_2}{CH}\!\!\sim\!\!\sim \\ & \longrightarrow & \\ \sim\!\!\sim\!\!CH_2\!-\!CH\!\!\sim\!\!\sim & & \sim\!\!\sim\!\!CH_2\!-\!CH\!\!\sim\!\!\sim \end{array} \qquad (8.54)$$

and

$$\sim\!\!\sim\!CH_2\!-\!\underset{\underset{\bullet}{O_2^{\bullet}}}{CH}\!\!\sim\!\!\sim \quad +H \longrightarrow \quad \sim\!\!\sim\!\!CH_2\!-\!\underset{\underset{H}{O_2}}{CH}\!\!\sim\!\!\sim \qquad (8.55)$$

As has been pointed out above, irradiation produces free radicals in solid polymers. In the solid the mobility of the free radicals is very restricted and some radicals may remain trapped for very long periods as they cannot encounter another radical with which to recombine. If radicals are formed in a crystalline region of a polymer, they are usually more firmly trapped.

In addition to the above chemical changes, irradiated polymers usually show physical changes. The mechanical properties may be affected, electrical conductivity may be induced in the material and very often colour changes occur. Most polymers turn yellow or brown with heavy irradiation, the dose required to produce this effect depending upon the structure of the polymer.

GRAFT POLYMERISATION

Graft polymerisation entails polymerising a monomer on to an already existing polymer so that a structure of the type

$$\begin{array}{c} -A\!-\!A\!-\!A\!-\!A\!-\!A\!-\!A\!-\!A\!-\!A\!-\!A\!-\!A\!- \\ | \\ B\!-\!B\!-\!B\!-\!B\!-\!B\!-\!B\!- \end{array}$$

is obtained. The polymer is irradiated so that the initiating radical and ionic sites are formed on it. This can be done in the presence of the monomer which has been swollen into the polymer. Unfortunately, irradiation in the presence of the monomer always leads to some homopolymerisation, which results simply in a mixture of two polymers instead of a graft. If, however, the polymer initially present is a solid, grafting is favoured, as the radical chains generated in the solid are less likely to terminate than those starting in the monomer. A more certain method of grafting is to irradiate the polymer before it is brought into contact with the monomer. Trapped radicals are produced which can subsequently initiate polymerisation when the monomer is introduced. For this to be successful, it is necessary to have radicals and ions on the polymer which are long-lived. A somewhat similar technique is to form a hydroperoxide or a peroxide on the polymer by irradiating it in the presence of oxygen. The peroxide can then be brought into contact with monomer and on heating the side chains should form. The reactions will depend on whether a peroxide or a hydroperoxide has been formed. With a peroxide,

$$
\begin{matrix} A \\ | \\ -O-O- \\ | \\ A \end{matrix} \quad \xrightarrow{heat} \quad \begin{matrix} A \\ | \\ -O^{\bullet} \\ | \\ A \end{matrix} + \begin{matrix} A \\ | \\ -O^{\bullet} \\ | \\ A \end{matrix} \qquad (8.56)
$$

$$
\begin{matrix} A \\ | \\ -O^{\bullet} \\ | \\ A \end{matrix} + nB \longrightarrow \begin{matrix} A \\ | \\ -O \sim\sim B \\ | \\ A \end{matrix} \qquad (8.57)
$$

or, if the peroxide is in the middle of a linear chain,

$$
A \sim O-O \sim A \xrightarrow{heat} 2A \sim O^{\bullet} \qquad (8.58)
$$

$$
A \sim O^{\bullet} + nB \longrightarrow A \sim O \sim B \qquad (8.59)
$$

With a hydroperoxide,

$$
\begin{matrix}
A \\
| \\
-O-OH \\
| \\
A
\end{matrix}
\quad \xrightarrow{\text{heat}} \quad
\begin{matrix}
A \\
| \\
-O^{\bullet} \\
| \\
A
\end{matrix}
\; + \; OH^{\bullet}
\qquad (8.60)
$$

The polymeric radical then initiates grafting polymerisation as in reaction (8.57) but the OH radical also initiates polymerisation of the monomer leading to the formation of some homopolymer.

If the hydroperoxide is located at the end of a linear chain, the radicals are formed according to

$$
A \sim O-OH \quad \xrightarrow{\text{heat}} \quad A \sim O^{\bullet} + OH^{\bullet} \qquad (8.61)
$$

and grafting follows according to reaction (8.59).

Whereas the properties of homopolymers formed by various initiating methods are practically the same, grafting alters the chemical and physical characteristics of the parent polymer drastically. Depending on the grafted material, there can be an increase or decrease in adhesion, ease of dyeing, solvent resistance, high-temperature dielectric properties, and ion exchange and tensile strength properties. Wood has been given quite different characteristics by combining it with various plastics.

HALOGEN COMPOUNDS

When other elements are present in organic molecules besides carbon and hydrogen, the number of reactions resulting from the absorption of ionising radiation may be reduced. The reason for this is a weak link which the heterogeneous atom may have with carbon or the weakening effect it may have on a neighbouring C—H or C—C bond. The organic halogen compounds are examples where a weaker than normal link is introduced into the molecule, while the alcohols, carbonyls and ketones represent compounds where the C—H links adjoining the carbonyl groups are weakened.

With the exception of fluorine, all other halogen-containing compounds show weaker carbon–halogen links than C—C or C—H bonds. The result is that in these compounds it is the halogen bond which breaks first on irradiation. Subsequent reactions, however, will vary because chlorine atoms can abstract hydrogen atoms from an organic molecule. Bromine atoms can do this only with difficulty and iodine atoms are completely inactive in this respect. Thus only a little HI ($G < 0.3$ molecules/100 eV) is produced when methyl and ethyl iodides are irradiated in the absence of oxygen. Consequently the yield of molecular iodine is higher.

Irradiation of the longer-chain bromides and chlorides can often result in a rearrangement of the atoms. This is due to recombination reactions, a typical one being the isomerisation of n-propyl chloride, which rearranges to the isopropyl chloride through a chain reaction ($G \approx 60$ molecules/100 eV). The radical propagation steps may be written

$$CH_3CH_2CH_2Cl + Cl \longrightarrow CH_3CHCH_2Cl + HCl \tag{8.62}$$

$$CH_3CHCH_2Cl \longrightarrow CH_3CHClCH_2 \tag{8.63}$$

$$CH_3CHClCH_2 + HCl \longrightarrow CH_3CHClCH_3 + Cl \tag{8.64}$$

A possible ionic mechanism is

$$CH_3CH_2CH_2Cl + e^- \longrightarrow CH_3CH_2\dot{C}H_2 + Cl^- \tag{8.65}$$

$$CH_3CH_2\dot{C}H_2 \longrightarrow CH_3\dot{C}HCH_3 \tag{8.66}$$

$$CH_3\dot{C}HCH_3 + CH_3CH_2CH_2Cl \longrightarrow CH_3CHClCH_3 + CH_3CH_2\dot{C}H_2 \tag{8.67}$$

The radical mechanism is preferred, as the yield increases somewhat on the addition of hydrogen chloride.

It has already been mentioned in Chapter 5 that any organic halogen compound generating HCl in the presence of a pH indicator can be used to show that a radiation dose of a certain level has been reached. Such an indication is useful on the side of containers holding sutures and syringes for radiation sterilisation.

Chain reactions with their high yields always make commercial exploitation a possibility, and so a considerable amount of work has been done on the production of sulphonyl chlorides, which are formed when sulphur dioxide, chlorine and a liquid hydrocarbon such as cyclohexane, pentane or dodecane are irradiated together. In this case G-values approach the order of 10^6 molecules/100 eV. With cyclohexane the propagation step will be

$$Cl + C_6H_{12} \longrightarrow HCl + C_6H_{11} \tag{8.68}$$

$$C_6H_{11} + SO_2 \longrightarrow C_6H_{11}SO_2 \tag{8.69}$$

$$C_6H_{11}SO_2 + Cl_2 \longrightarrow C_6H_{11}SO_2Cl + Cl \tag{8.70}$$

The reaction can be carried out effectively with any large radiation source. A similar reaction with a high yield ($G \approx 10^5$ molecules/100 eV) and based on a parallel photochemical mechanism is the radiation chemical formation of the weedkiller gamma benzene hexachloride:

$$C_6H_6 + 3Cl_2 \longrightarrow C_6H_6Cl_6 \qquad (8.71)$$

Toluene can be chlorinated in a similar fashion. Again, in keeping with photochemical reactions, ionising radiation can be used to sulphonate hydrocarbons. Here, oxygen replaces the chlorine molecule and therefore the mixture to be irradiated contains the hydrocarbon, sulphur dioxide and oxygen. In comparison with the photo-chemical process, ionising radiations are preferred because they can penetrate in depth to a much greater extent than can photons. Despite the fact that ethyl bromide has been made in a development plant, none of the above products has a sufficiently large market to justify its full-scale production.

ALCOHOLS

Alcohols, ethers, acids, esters and carbonyl compounds all belong to the type of substance containing a group which weakens adjoining bonds. Also, as they contain at least one oxygen atom, the general radiation chemistry of these compounds is a mixture of effects as between the polar and non-polar media described in *Table 8.1*. Formation of the products is due to both 'molecular' and 'radical' processes. The latter are, again, capable of being altered by the introduction of radical scavengers to the system.

The irradiation of methanol, for example, produces ion and radical intermediates of the type CH_3OH^+, CH_2OH^+, CHO^+, $CH_3OH_2^+$, CH_3^+, CH_3OH^-, CH_3O^- and H, CH_2OH, CH_3, OH, respectively. Hydrogen gas, formaldehyde and ethylene glycol are the main products of irradiation.

Irradiation of ethanol results in the formation of analogous ions and radicals, which lead to the final main products of hydrogen gas, acetaldehyde, 2,3-butanediol and, with much lower yields, methane and carbon monoxide. In the presence of oxygen, practically no 2, 3-butanediol is formed and the yield of acetaldehyde is doubled. In conjunction with these effects, the yield of hydrogen drops markedly and substantial quantities of hydrogen peroxide appear. All this is consistent with the formation of organic peroxy and hydroperoxy radicals (analogous to the behaviour of water) which prevent radical recombination.

From the protective effect of the benzene ring it would be expected that benzyl alcohol would give very low radiation yields of benzaldehyde, even in the presence of oxygen. Actually, yields of aldehyde and peroxide with G-values of about 50 molecules/100 eV are found. A chain mechanism involving a peroxy radical is thought to be responsible for this effect, the propagation steps being

$$C_6H_5-\dot{C}HOH + O_2 \longrightarrow C_6H_5-\underset{O_2^{\bullet}}{CHOH} \qquad (8.72)$$

$$C_6H_5-\underset{O_2^{\bullet}}{CHOH} + C_6H_5-CH_2OH \longrightarrow C_6H_5-\dot{C}HOH + C_6H_5-\underset{O_2H}{CHOH} \qquad (8.73)$$

$$C_6H_5-\underset{O_2H}{CHOH} \longrightarrow C_6H_5-CHO + H_2O_2 \qquad (8.74)$$

In fact, the benzene ring still acts in a protecting manner, but only with respect to the initial formation of the radical $C_6H_5-\dot{C}HOH$. This can be shown by carrying out the irradiation in the absence of oxygen.

ACIDS, ETHERS, KETONES AND ESTERS

Irradiation of organic acids results in decarboxylation and hydrocarbon formation. From acetic acid, for example, carbon dioxide and methane are the main products. Obviously, the molecule breaks in at least three different places:

$$CH_3 \bcancel{+} \overset{O}{\underset{\parallel}{C}} \bcancel{+} O \bcancel{+} H$$

When other heterogeneous atoms besides oxygen are present in the organic molecule, then other radiation products besides carbon

N

dioxide and methane may be produced. Thus amino acids yield mainly ammonia, and hydrogen sulphide can be expected from acids containing sulphur atoms.

The effect of ionising radiation on acids has been studied mainly in aqueous solution and it was here, from the irradiation of monochloracetic acid, that one of the first indications of the occurrence of the solvated electron came.

Ethers, ketones and esters also fragment on irradiation, giving rise mainly to methane, hydrogen, carbon dioxide and carbon monoxide among a variety of products. Once again, if these substances are irradiated in aqueous solution, the number of products is reduced, as only three initiating species occur: OH, H and e_{aq}^-.

The irradiation of organic substances of biological interest in the presence and absence of water will be discussed in the next chapter.

9

Aspects of biological systems

There are two areas in which the effects of ionising radiation have been intensively studied. One is their effect on materials used in nuclear reactors and allied industries. The basis for this has been covered in the chapters on gases, solids and liquids. The other area is the effect of ionising radiation on biological systems. In this case great interest has been shown for the following reasons:

(1) High doses of radiation can kill (accidents and nuclear weapons).

(2) Cancer of various kinds can be induced by ionising rays (early radiologists and users of radium paint).

(3) The frequency of genetic variations can be increased (animal experiments).

(4) Cancer cells can be destroyed preferentially (radiotherapy).

(5) The destruction of antibodies by ionising radiations may reduce rejection in organ transplantation.

All these are general statements on man as a biological organism. What has become apparent since radiation chemists tried to 'help' radiation medicine and interpret biological effects (which occupy from minutes to years) is that the simple transfer of result and theory from physics and chemistry (where the effects occupy less than 10^{-6} s) to biology is not possible. It is therefore of importance that the radiation chemist have some understanding of those biological effects which have no parallel in chemical systems.

The first of these is 'biological recovery'. In any biological system which consists of a large number of cells some die and are replaced by new ones. Ionising radiation kills cells and their successful replacement is the biological recovery. The organism may not be able to replace every type of cell, but, in general, those most sensitive to radiation are most easily replaced. When too many cells have been killed at once and replacement is impossible, death of the animal as a whole results. Radiation death of the organism is therefore dependent on whether the same total radiation is given all at once in 'acute' form (within 24 h), or in 'chronic' form spread over days, weeks and years. The dose may also be given in 'fractionated' form, when small acute doses are given at intervals. The total dose which can be given in a chronic or fractionated manner before the death of the organism occurs can be up to three times larger than the acute lethal dose.

In man an acute dose given (by X-rays or γ-rays) over the whole body has the following effects.

(1) Up to 100 rem: no noticeable effect in healthy people. Sick people are affected by as little as 25 rem. (See definition of rem in Chapter 5.)

(2) 100–300 rem: radiation sickness for a few days (caused by the death of cells and their removal), nervous or permeability effects. Fall in white blood cell count with recovery apparent after a week.

(3) 300–1000 rem: fall in white blood cell count, haemorrhages (occurring from fifth day to third week depending on the dose), secondary infections (third week) with little chance of recovery at doses above 600 rem.

Where recovery does take place, it may be followed by delayed effects due to ionising radiation (see below). These include anaemia, fibrosis, dermatitis, cancer, leukaemia, life-shortening, cataract and an increase in hereditary abnormalities.

The biologist describes the acute total body dose at which the life of a man is in the balance as the $LD_{50/30}$. This is the dose which will be lethal to 50% of the population within 30 days of absorption. In man the $LD_{50/30}$ is approximately 400 ± 100 rem. In dogs, rats and adult fruit-flies the $LD_{50/30}$ is 300 rad, 900 rad and 60 000 rad, respectively. In bacteria it is between 500 and 10^5 rad depending on the stage of development of the bacteria (i.e. whether it is active or dormant).

An acute whole-body dose to man above 1000 rem results in death within three to five days from damage to the intestinal tract, with diarrhoea and the loss of fluids and blood. When only a small volume of the body is exposed to radiation, as in the radiation therapy of cancer, the doses which may be given can be appreciably higher than the $LD_{50/30}$. Cancer cells grow and divide more rapidly than normal cells and are therefore more sensitive to ionising radiation. High localised doses are necessary if all the cancer cells present are to be killed. Variations in radiation sensitivity exist not only between cancer and normal cells, but also between the different types of cells found in the body. Again, the smaller the specialisation of the cell, the more rapidly it grows and divides and the more sensitive it is to ionising radiation (Law of Bergonie and Tribondeau). Thus in the body there are:

(1) very sensitive cells, e.g. sperm-forming, blood-forming and lymph-forming, growing tissues such as embryos, ovaries and thyroid;

(2) medium-sensitive cells, e.g. skin, gastrointestinal tract;

(3) least sensitive cells, e.g. bone, connective tissue, liver, kidney, nerve, brain, muscle, thyroid gland.

If the small volume irradiated in therapy includes a substantial proportion of very sensitive cells, then many of the effects seen in whole-body exposure will also be observed under these conditions.

The radiation source might not be external but could arise from radioactive isotopes absorbed in the body. Here, also, preferential uptake in a certain organ may result in a high localised dose. The therapeutic use of iodine-131 for cancer of the thyroid is a case in point, as is the deleterious effect of strontium/yttrium-90, which accumulates in bone near the blood-forming cells.

Evidence in man for the delayed effects resulting from the absorption of ionising radiation has come over the years following the discovery and use of these rays. These effects would appear some time following the absorption of acute doses of the $LD_{50/30}$ or chronic and fractionated doses of approximately three times the $LD_{50/30}$. Bad working conditions made it possible for the early radiologists to receive up to 20 rem/d on their hands. As a result, they had dermatitis of the hands five times more frequently than did the rest of the population. They also developed cancer; 100 such workers died in 1922 alone. People exposed to radiation from the atomic bomb at Hiroshima and Nagasaki also showed a higher frequency (twice normal) of cancer and particularly of leukaemia (50 times normal), although with the latter the total number of cases was small. On the other hand, workers who had used radium paint showed no increase in the frequency of leukaemia, but an increase of cancer. The formation of radiation cataracts has been observed in personnel working with cyclotrons, neutrons and X-rays above 200 rem, and following radium needle therapy in cancer of the tongue.

The induction of cancer occurs at an optimum dose which is

sufficiently high to affect the cells, yet not high enough to kill them all. This optimum dose is higher for young than old animals. For example, in one-day-old mice it is 800 rad and in older mice it is 200 rad. This suggests that at least two events are necessary for cancer induction, the non-irradiation factor increasing with age. At present the mechanism by which normal cells change to neoplasms is not fully understood. It follows that the way in which sublethal doses of ionising radiation induce cancer is also obscure. It is known, however, that radiation affects the mutation rate and thus affects DNA (deoxyribonucleic acid) and associated large molecules (see below). The daughter cells may not have the same characteristics and differentiative properties, and may be converted into neoplastic cells. The more such conversions occur, the more likelihood of the cancer taking hold. On the other hand, irradiation may weaken the external and nuclear membranes of the cell and change the enzyme balance in the cell. This could lead to a change in the rate of meta-bolism, reproduction, immediate environment and, finally, differen-tiative properties. Again, weakening of the membranes might permit viruses which would otherwise lie dormant to become active. Although whole armies of researchers are working in this field, no solution is yet in sight. From the chemist's point of view, however, it makes the study of the effects of radiation on molecules and structures of biological importance of the greatest interest.

Another delayed effect, at doses greater than 50 rad, which has been shown in mice is a shortening of life-span. It may well be that cells which do not divide, or divide slowly, accumulate damage from chronic radiation. As with all delayed effects, it is impossible to say whether there is a minimum dose below which these effects are not induced.

A delayed effect not so far discussed is one resulting from the irradiation of the reproductive organs. This can take three forms, two resulting from irradiation prior to fertilisation and one from the exposure of the foetus in the uterus. Radiation of the reproductive organs can lead to sterilisation or partial sterilisation. In the latter case only sperm cells in a particular state of development are killed and fertility returns in the male as the reproductive cells produce more sperm. From animal experiments, it has been extrapolated that 400 and 600 rem will sterilise the ovaries and testes, respectively. Approximately 250 rem may give rise to partial sterilisation.

Genetic effects make themselves felt through the transfer of pattern (i.e. the thousands of inherited characteristics). This applies to cell reproduction (somatic inheritance) as well as the reproduction of the whole animal (genetic inheritance). In both cases the hereditary pattern is carried by the chromosomes situated in the nuclei of living cells. Within each chromosome very large molecules (relative molecular mass c. 10^6) of DNA and protein molecules are arranged in a certain order. Alteration of this order can produce variations in the inherited characteristics of the offspring, be it a cell or animal. Each DNA molecule itself has a pattern made up of a sequence of four bases (adenine, thymine, cytosine and guanine) joined together by a sugar (deoxyribose) phosphate ester chain. The chemical compositions and configurations of the DNA and protein molecules are known as their primary structure, but the special conformation (helix) geometry of these very large molecules also affects their biological properties and is known as their secondary structure. Proteins and nucleoproteins have also tertiary structural properties, because they can often combine into even larger definite 'molecules' with specific biological functions. RNA (ribonucleic acid), although similar in primary structure to DNA, is generally a much smaller molecule with a variety of functions in the cell. One of these is the synthesis of protein and enzyme molecules. RNA does not appear to be as biologically sensitive to ionising radiation as is DNA.

When cells divide, DNA–protein patterns are transferred to daughter cells with remarkable accuracy. Only about once in a hundred thousand times does a change in pattern occur. This change in pattern is known as a mutation. In ordinary cell division (mitosis) transfer of the pattern is brought about by a doubling of the DNA molecules with subsequent halving when the cell divides. In the case of reproductive cells halving occurs first and is followed by restitution to the normal level of chromosomes by fertilisation (meiosis). Simple variations in the reproductive cells can affect the next or many later generations. The general effect of ionising radiation is to increase the mutation rate of each known mutation. New mutations are not found, but cannot be entirely discounted. This information has been obtained from experiments with mice and fruit-flies. As the genetic mutation process is fundamentally the same in all animals, it can be expected that radiation exposure of the reproductive organs of man will also lead to an increase in the number of carriers of known inheritable diseases. These would include recessive variations such as albinism and phenylketonuria. Dominant variations such as mongolism and granulocytic leukaemia would be eliminated after one generation but recessive variations would continue. It is only recently that man has subjected himself to radiation doses larger than background. Actually, medical uses of radiation and fall-out from nuclear tests have increased background radiation only by about 20%. Even if the background were to double, the statistical variation in the mutation rate would mask the small increase in mutation rate expected from such a dose (3·6 rem in 30 years; one generation).

Once fertilisation has taken place, the growing foetus in the uterus becomes extremely sensitive to radiation. This is because its cells are dividing and differentiating very rapidly. In the human, the most sensitive period appears to be up to the 38th day of pregnancy. An acute dose of 40 rem on the 28th day is considered dangerous with respect to the formation of abnormalities. Exposure in late pregnancy increases slightly the small chance of leukaemia and cancer. Irradiation *in utero* should thus be kept to a minimum.

If cell death itself is now considered (from experiments in which the cells are counted), many radiation biologists think that the cell contains a small sensitive volume which, if damaged by ionising radiation, will cause the cell either to die at once (interphase death) or to die after only a few divisions (reproductive death, dealt with above). Any further radiation absorbed in the cell will be superfluous as far as the cell death is concerned. As to the most sensitive volume, this has been called the 'target' and has led to the 'target theory', which states that if the target is hit, the cell dies. The exact nature of this target has not been established and probably varies with the age of the cell and the type of cell. Obvious candidates for the target, which should be highly sensitive and unique, are the chromosomes in the nucleus, the nuclear membrane or the lysosomes in the cell cytoplasm. Rupture of the latter has been shown to release hydrolytic enzymes which can break down other cell constituents. Apart from complete cell death, it has also been observed that the absorption of radiation can lead to an increase in metabolic activity as well as to a delay in cell division. Both of these observations are indicative of the recovery process mentioned above.

The various findings in radiation biology have pointed to the sensitivity to irradiation of the cell nucleus and the reactions stemming from it. As a result, DNA and its associated proteins, RNA and the hydrolytic enzymes have been the subjects of a great deal of study by radiation chemists. They are just as limited, however, as are other cell chemists by the experimental difficulties and the lack of methods available when working with very large molecules. As a result, it has been virtually impossible to carry out experiments which link directly the radiation effects on cell mutation, growth and death with observations of changes in the corresponding DNA and protein molecules. If viscosity is taken as a criterion of DNA size, then the decrease in viscosity observed on irradiation of DNA indicates chain fracture. This breaking of the DNA molecule by ionising radiation could certainly account for the chromosomal aberrations and disjunctions which can be observed under the electron microscope. Where double DNA strands are concerned, a double chain fracture could lead to death on cell division. Recombination or cross-linking of DNA strands could lead to a viable yet changed molecule giving rise to mutations. The difficulty with the viscosity measurements lies in the fact that the DNA and its associated molecules have to be removed from the cell nucleus. From a biological point of view, this treatment is severe in that membranes have to be broken, enzymes are released and the electrochemical environment is changed. Working up of the substances for analysis can thus change them. How accurately viscosity and sedimentation changes mirror the radiation effects in the nucleus of the cell is difficult to say. Electron spin resonance studies have shown that free radicals are formed in biological systems as a result of the absorption of radiation. Some of the effects of free radicals on components of DNA and proteins are known. The loss of amino groups from the bases in DNA and the amino acids which make up proteins has been well documented. This deamination is also observed when DNA itself is irradiated outside the cell (*in vitro*). Another effect which has been measured is the release of free phosphate groups resulting from radiation-induced breakage of the phosphodiester bonds. All these effects are substantially diminished if the DNA is irradiated in the presence of its associated nucleoprotein. Apparently the protein sheath acts in a protective manner through competition for radicals and by providing steric hindrance. A parallel can be seen in the biological effects of radiation, because living cells are particularly sensitive to radiation at certain stages during their life-cycle, presumably when no protein is associated with the DNA molecule. Nucleoproteins irradiated at concentrations found in the cell show changes in their chromatographic elution pattern. It is reasonable to assume that any radiation attack they suffer will be similar to that found with non-nucleoproteins. Here breakage of the peptide bond is seldom observed (with the exception of isoleucine) and the points of attack depend on the nature of the amino acid side-chain. Sulphur amino acids appear to be the most sensitive, oxidising easily to their disulphide form when possible. In aqueous systems solvated electrons add on to any CO_2 present, which gives rise to the radical CO_2^-. Carbon dioxide labelled with carbon-14 has been shown to add on to the minute lipid fraction in the nucleus rather than anything else, so perhaps the ionisation processes which take place in the DNA complex overshadow the indirect radical mechanisms.

Most radiation biological effects are markedly enhanced by going from anoxic (absence of oxygen) to aerobic (presence of air) conditions. This shows that oxygen is a radiation sensitiser and that irreversible organoperoxy radicals are active here. Any increase in oxygen tension beyond aeration has little effect. Although tissues are normally in equilibrium with air there are biological situations (e.g.

in certain tumours) where, because of multicompartmentalisation and encapsulation, metabolic processes occur under anoxic conditions. In these cases the application of oxygen under pressure will increase the effects produced by ionising radiation, a result which has now been taken into the therapeutic field. Other sensitisers have been investigated, to increase the efficiency of therapy and to pinpoint specific biological loci. A typical cell sensitiser is 5-bromouracil. It has been shown that this compound replaces thymine in DNA molecules, and this is therefore another pointer to the great radiation sensitivity which this molecule has in the cell life. The reverse of sensitisation to ionising radiation is chemical protection. Here it must be said at the outset that there are no real protectors against ionising radiation. The only real protection is to get as far away from the source as possible and, failing that, to use a solid shield until the radiation dose is near to that of the background level. Some experiments have been carried out with animals and bacteria in which a 'protective agent', such as cysteine or cystamine, was given immediately prior to exposure. These substances were observed to depress the sensitivity of the cells towards ionising radiation (i.e. a higher dose had to be given to achieve the same radiation biological damage). The precise function of compounds which have been found to be radiation chemical protectors appears to lie in a combination of reactions. These include the removal of free radicals before they can attack radiation-sensitive molecules and the removal of oxygen in the immediate neighbourhood of the radiation track. The latter reaction prevents the formation of organoperoxy compounds and reduces the number of irreversible changes. Finally, the radiation protective compounds may complex with particular sensitive sites on large molecules, so these can function again after the ions and radicals have disappeared.

A biochemical mechanism which was examined at an early stage by radiation chemists was the deactivation of enzymes. This helped to confirm the importance of the indirect effect observed in solution. Although deactivation of enzymes will retard metabolic and possibly mitotic processes, it is difficult to envisage the whole of an existing group of enzymes being destroyed by radiation so that the cell can no longer function. Far more serious is the damage to DNA and its associated molecules which would completely prevent the synthesis of an essential enzyme. Also lethal would be the release of enzymes, through membrane-rupture, into sections of the cell where they can break down vital molecules. A change in the permeability characteristics of membranes might well result from the passage of an ionising particle through lipid–protein or lipid–cellulose structures. The formation of peroxy and carboxyl groups along fatty acid chains would certainly change their hydrophobic properties. Thus it comes about that divalent metal gegen-ions show some protective effect in that they help to retain the electronic structure of the membrane.

Both membranes and DNA–protein complexes are semi-solid, semiconductor-like structures and, as such, will be particularly sensitive to ionisations concentrated into a small volume. It is therefore not surprising that α-particles have a greater biological effect than has the same dose of the less densely ionising X-rays or γ-rays. Indeed, for most biological effects a direct relationship is observed between LET and the Relative Biological Effectiveness (RBE) of a particular type of radiation, e.g.:

$$\text{RBE for } \alpha = \left(\begin{matrix}\text{Dose required to produce}\\ \text{a certain biological effect}\\ \text{with X-rays or } \gamma\text{-rays}\end{matrix}\right) \Bigg/ \left(\begin{matrix}\text{Dose required to produce}\\ \text{the same biological effect}\\ \text{with } \alpha\text{-particles}\end{matrix}\right)$$

$$\approx 10$$

Many experiments have been undertaken with the wide variety of substances found in the living cell. They have been examined in the dry state and in solution. Their sensitivity to the primary radiation products has been accurately measured by means of pulse radiolysis and, where possible, their radiation breakdown products have been analysed, the yields determined and mechanisms proposed. It has even been shown that some of the molecules (e.g. amino acids) can be synthesised with the help of ionising radiation. All of these very interesting scientific studies, however, have not been able to give complete and satisfactory answers to the five interactions between man and radiation mentioned at the beginning of this chapter. Such answers will probably have to wait for further elucidation of cell chemistry and the physicochemical behaviour of the very large molecules. This in turn will depend on the development of better experimental techniques.

Bibliography

This bibliography is not intended as a comprehensive list but selected references are given to indicate areas for the next stage of study.

BOOKS

An Introduction to Radiation Chemistry, J. W. T. Spinks and R. J. Woods, Wiley, New York, 1964

Principles of Radiation Chemistry, J. H. O'Donnell and D. F. Sangster, Arnold, London, 1969

Introduction to Radiation Chemistry, I. V. Vereshchinsky and A. K. Pikaev, Davey, New York, 1965

Radiation Chemistry of Organic Compounds, A. J. Swallow, Pergamon, Oxford, 1960

Fundamentals of Radiobiology, Z. M. Baeg and P. Alexander, 2nd edn, Pergamon, Oxford, 1961

Radiation Protection – I.C.R.P. Publications, Pergamon

Radiation Chemistry of Polymeric Systems, A. Chapiro, Vol. XV in the series, *Chemistry of High Polymers*, Interscience, New York, 1962

'Electrochemical Processes in Glow Discharge at the Gas-Solution Interface', A. Hickling, in *Modern Aspects of Electrochemistry*, Vol. 6, p. 325, Butterworths, London, 1971

COLLECTED PAPERS AND REVIEWS

Fundamental Processes in Radiation Chemistry, P. Ausloos (Ed.), Wiley, New York, 1968

Energy Transfer in Radiation Processes, G. O. Phillips (Ed.), Elsevier, Amsterdam, 1966

Pulse Radiolysis, M. Ebert, J. P. Keene, A. J. Swallow and J. H. Baxendale (Eds), Academic Press, London, 1965

Pulse Radiolysis, M. S. Matheson and L. M. Dorfman, MIT Press, Cambridge, Mass., 1969

Solvated Electron, E. J. Hart (Ed.), Advances in Chemistry Series No. 50, American Chemical Society, 1965

The Chemistry of Ionization and Excitation, G. R. A. Johnson and G. Scholes (Eds), Taylor and Francis, London, 1967

Radiation Chemistry of Aqueous Systems, G. Stein (Ed.), Wiley, London, 1968

Index

Absorbed dose, of radiation, 85–89
Absorption coefficient
  atomic, 46, 47
  electronic, 46, 47
  energy Compton, 49
  linear, 46
  mass, 46, 47, 86, 87
  total Compton, 49
Abstraction reactions
  of excited states, 72
  of free radicals, 78, 140, 141
  of hydrogen atoms, 140
  of hydroxyl radicals, 141
Accelerating tube, 23, 24
Acceptor levels, 163
Acetylene, 156, 174
Acids, 185, 186
Activity, 15, 96
Acute dose, of radiation, 188, 189
Acute total body dose, of radiation,
  188
Addition reactions
  of electrons, 137
  of excited states, 72
  of free radicals, 78, 140, 142
  of hydrogen atoms, 140
  of hydroxyl radicals, 142
Air-dose, 88
Alcohols, 184, 185
Alkali halide crystals, 160

Alpha-particle tracks, 37
Alpha-rays, 2, 10
Alpha-sources, 20, 21
Amino acids, 186
Annihilation radiation, 50
Atomic absorption coefficient, 46, 47
Attenuation, 45
Auger electrons, 48

Barrier detectors, 165
γ-Benzene hexachloride, 184
Bergonie and Tribondeau's law, 189
Beta-rays, 2, 10
Beta-sources, 19, 20
Betatron, 25
Biological effects, 11
Biological recovery, 187
Bragg curve, 26
Bragg–Gray cavity principle, 87, 97
Bremsstrahlung, 22, 33, 36
Bromide, 124, 125
5-Bromouracil, 194

Cage effect, 55, 71, 169
Cancer, 187–189
Capture
  of electrons, 60
  of neutrons, 43, 51
Carbon dioxide, 152, 153, 173
Carbon monoxide, 173

Castel key system, 19
Cathode fall, 28, 29
Cathode rays, 8
Cavity-dose, 88
Cell death, 192
Cerenkov radiation, 76
Ceric sulphate, 121–123
Ceric sulphate dosimeter, 92
CGS units, 4
Chain reaction, 149, 183
Chain scission, 179
Characteristic X-rays, 23, 47
Charge transfer reactions, 59
    of hydrogen atoms, 140
Charged particles, 31–40
    energy loss by, 35, 36
    range of, 36–40
Chemical dosimeters, 89–93
Chemical protection, 194
Chromosomes, 190, 192
Chronic dose, 188
Cis–trans isomerisation, 71, 175
Cloud chamber, 57
Cobalt-60, 15
Collisions, 31
    elastic, 31, 34, 35
    inelastic, 32, 34
Colour centres, 160–163
Combination reactions, 78
Compton absorption coefficient, 49
Compton effect, 48, 49, 51
Condenser-r-chambers, 82–84
Conductance, 163–165
Conduction bands, 163, 164
Conductors, 163, 164
Cross-linking, 172, 177, 178
Cross-sections, 43, 44, 50, 157
Cyclohexane, 172, 173
Cystamine, 174, 194
Cysteine, 194

Deactivation of enzymes, 194
Deamination, 140, 193
Decarboxylation, 185
Decay constant, 96
Defects of diffusion model, 142–144
Degradation, 177, 179
Delta-rays, 53, 55
Deposition of energy, 36
Diffusion model, 101, 103
    defects of, 142–144
    modifications to, 144–146

2,2-Dimethyl butane, 172
2,5-Diphenyloxazole, 174
Diradicals, 79
Direct effect, definition of, 3
Displacements, 157, 159, 165
Disproportionation reactions, 139
Dissociation
    of excited states, 68–71
    of free radicals, 77, 138
DNA, 191–195
Dodecane, 172
Dominant variations, 191
Donor levels, 163
Dose
    absorbed, 85–89
    acute, 188, 189
    acute total body, 188
    air, 88
    cavity, 88
    chronic, 188
    exposure, 85–88
    fractionated, 188
    lethal, 188
    personal, 95
Dose rate, 84
Dosimeters
    ceric sulphate, 92
    Fricke, 89–91
    solid, 93
    thermoluminescent, 93
    thiocyanate, 92
Dosimetry
    chemical, 89–93
    ionisation, 82–88
    of internal sources, 95, 96
    neutron, 97
Dry electrons, 77, 145

Effects of ionising radiations, 51–55
Elastic collisions, 31, 157
    of charged particles, 34, 35
Elastic scattering, 34
    of alpha-particles, 37
    of neutrons, 40–42, 51
Electrical discharges, 2
Electromagnetic radiation, 2, 31, 45–50
Electron density, 34
Electron paramagnetic (or spin) resonance spectroscopy, 73, 75, 168
Electron transfer reactions, 71, 72, 79, 142

Electrons
  reactions of, 60, 61
  subexcitation, 61
  solvation of, 61
Electronic absorption coefficient, 46, 47
Emission of radiation, 33
Energy loss of charged particles, 35, 36
Esters, 185, 186
Ethane, 154
Ethanol, 184
Ethers, 185, 186
Ethyl bromide, 184
Excited states, 34
  conversion of, to ground state, 65–68
  detection and study of, 61, 62
  reactions of, 62–72
    bimolecular, 71–72
    unimolecular, 68–71
Exposure dose, of radiation, 85–88
External sources, of ionising radiations, 13–26
  alpha-, 20, 21
  beta-, 19, 20
  gamma-, 15–19
  machines, 21–26
  nuclear reactors, 21
  radioactive isotopes, 13–21
  spent reactor fuel, 21
Extrapolated range
  of alpha-particles, 37
  of beta-particles, 39
  of electrons, 39

Fast neutrons, 40, 42
F-centres, 161, 162
Ferrocyanide and ferricyanide, 127, 128
Ferrous sulphate, 119–121
Ferrous sulphate–cupric sulphate system, 123, 124
Film badge, 95
Fission products, 20, 165
Flash photolysis, 75
Fluorescence, 65
Fractionated dose, 188
Franck–Condon principle, 62, 65
Franck–Rabinowitch effect, 71
Free radicals, 56
  detection and study of, 73–77

reactions of, 77–79
  bimolecular, 78, 79
  unimolecular, 77, 78
Fricke dosimeter, 89–92
Frozen solutions, 163

Gamma-rays, 2, 10
Gamma-sources, 15–19
Gases, 147–156
Gel formation, 172, 177
Genetic effects of radiation, 187, 190
Glow discharge electrolysis, 28–30
Graft polymerisation, 180–182
Graphite moderator, 165
G-values, 80, 117
  $G_{H_2O_2}$, determination of, 122, 123, 126–128
  G_{H_2}, determination of, 122, 125, 127, 128
  G_{OH}, determination of, 122, 126, 128, 129
  G_H, determination of, 121, 126, 127, 129
  G_{e^-}, determination of, 121, 126, 127, 129
  G_{HO_2}, determination of, 124

Half-life, definition of, 15
Half-value thickness, 11
Halogen compounds, 182–184
Health hazards, 16
Heterogeneous catalysts, 168
Hexane, 171
Highly excited state, 58
High-pressure mass spectrometry, 58
Homopolymerisation, 181
Hot radicals, 59, 70
Hot spots, 157
Hydrated electron, 92
  reactions of adducts of, 138, 139
Hydrocarbons
  gaseous, 153–156
  liquid, 171–175
    straight chain, 172
Hydrogen–chlorine system, 149, 150
Hydrogen–deuterium system, 148, 149
Hydrogen atoms, 99
  reactions of, 139–141
Hydroperoxides, 173, 180, 181
Hydroperoxy radicals, 184

Hydroxyl radicals, 99, 104
  reactions of, 141–142

Impurity semiconductor, 163, 164
Indirect effect, definition of, 3
Inelastic collisions, 32, 157
  of charged particles, 33, 34, 151
Inelastic scattering of neutrons, 42, 51
Initiation of chemical reactions, 1–4
Internal conversion, to ground state, 65, 66
Internal radiation hazard, 20
Internal sources, 27
Interphase death, 192
Intersystem crossing, 67, 68
Inter-track reactions, 135
Intra-track reactions, 134
Intrinsic semiconductor, 163–165
Iodine-131 therapy, 189
Ion–molecule reactions, 57, 59, 171
Ion pair, 36, 52
  energy of formation of, 52, 147
Ionic yield, 80
Ionisation chamber, 9, 57, 82, 84, 85, 165
  air-equivalent, 85
  thimble, 85, 87
Ionisation dosimetry, 82–88
Ions
  detection and study of, 57, 58
  reactions of, 58–60
Isobutylene, 177

Ketones, 185, 186
Kitchen and Pease model, 160
Knock-ons, 157

Labyrinth source, 17–19
LD$_{50/30}$, 188
LET, 54, 55, 104
Lethal dose, of radiation, 188
Linear absorption coefficient, 46
Linear accelerator, 25
Linear energy transfer, see LET

Machine sources, of ionising radiations, 21–26
Mass absorption coefficient, 46, 47, 86, 87
Mass spectrometer, 57
Mass stopping power, 86
Material balance equations, 117, 118
Maximum permissible level, 16
Maximum range, 10, 38

Mean range, 37, 38
Measurement of low doses, 93–95
Meiosis, 191
Metabolic activity, 192
Metal filters, 23
Metals, effect of radiation on, 164
Methane, 153
Methanol, 184
Methyl methacrylate, 175
Mitosis, 191
MKS units, 4
Molecular products, 107
Molecular yields, 106–109
  determination of, 118–130
    in acid solutions, 119–124
    in alkaline solutions, 127, 128
    in neutral solutions, 124–126
  effect of dose rate on, 134–136
  effect of LET on, 130–134
  effect of pH on, 134
  limitations of, 114–116
  variation of, with concentration, 115
Multiplicity, 64
Mutation rate, 190, 191

Negative ion vacancy, 160
Neutral particles, 31
Neutralisation of ions, 58
Neutron dosimetry, 97
Neutrons, 2, 40–44
Nitrogen–oxygen system, 151
Nitrous oxide, 152
Non-conductors, 163–165
Non-hydrated electrons, 145
Non-polar media, 170
Non-radiative energy transfer, 68
Nuclear fission reactions, 20
Nuclear fuel, 165
Nuclear reactions, 15, 42, 51
Nuclear reactors, 21, 16, 187
Nuclear tests, 191

Origin of radiation chemistry, 8–12
Oxygen, 150
  and polymers, 179
  in aqueous solution, 125, 126
  reaction of free radicals with, 79
Oxygen–hydrogen system, in aqueous solution, 126

Pair production, 49–51
Peptide bond, 193

Perhydroxyl radical, 99, 105
Peroxides, 173, 180, 181
Peroxy radicals, 79, 173, 184
Personal dose, of radiation, 95
Phosphorescence, 67, 68
Photochemistry, 3
Photoelectric effect, 47, 48, 51
Pile unit, definition of, 21
p–n junctions, 165
Polar media, 170
Polymerisation, 175–178
  free radical, 177
  graft, 180–182
  ionic, 177
  solid state, 177
Polymers, irradiation of, 178–180
Positive holes, 165, 167
Positive ion vacancy, 161
Positrons, 12, 50
Practical range, 39
Predissociation, 69
Primary displacements, 157
Primary product yields, 130
Primary products, 108
Primary structure, 191
Propane, 154–156
n-Propyl chloride, 183
Protection, radiation, 174
Pulse radiolysis, 27, 75, 92, 129, 130,
  135

Quality factor, 94

Rad, definition of, 86
Radiation chemical yields, 79–81
  expression of, as G-values, 80, 81
  ionic, 80
Radiation protective effect, 174, 194
Radiation resistance, 173
Radiation sensitiser, 193
Radiation sensitivity, 189
Radiation sickness, 188
Radiation sterilisation, 183
Radiation vessels, 27, 28
Radiative attachment, 60
Radiative conversion, 65
Radical combination reactions, 78,
  141, 142
Radical reaction rate constants, 136,
  137
Radical yields, 106–109
  determination of, 118–130
    in acid solutions, 119–124

  in alkaline solutions, 127, 128
  in neutral solutions, 124–126
  effect of dose rate on, 134–136
  effect of LET on, 130–134
  effect of pH on, 134
  limitations of, 114–116
Radioactivity, 9
Radiolysis
  of aqueous solutions, 116–118
  of pure water, 109–114
Radium needle therapy, 189
Range, 10
  of alpha-particles, 37, 38
  of beta-particles, 39
  of charged particles, 36–40
  extrapolated, 37–39
  maximum, 10, 38
  mean, 37, 38
  of monoenergetic electrons, 39
  practical, 39
R-centre, 161, 162
Reactions
  of hydrated electron adducts, 137–
    139
  of hydrogen atoms, 139, 140
  of hydroxyl radicals, 141, 142
Rearrangement of free radicals, 77
Recessive variations, 191
Recoil energy, 42, 48
Recoil reactions, 20, 27, 48, 97
Relative biological effectiveness, 195
Rem, definition of, 94
RNA, 191
Reproductive death, 192
Repulsive states, 69, 70
Resonance energy, 43
Resonance process, 60
Röntgen, definition of, 82
Rubbers, effect of radiation on, 179

Scattering, 32, 34, 35
Scattering coefficient, 49
Scavenger, 106
Scintillating phosphor, 174
Secondary structure, 191
Selection rule, 64
Semiconductors, 163, 164
Shock waves, 1
SI units, 4
Singlet states, 64
Slow electrons, 56, 64
Slow neutrons, 42, 43
Solid dosimeters, 93

Solid polymers, 178–180
Solids, 156–168
    chemical changes in, 166–168
    dimensional changes in, 165, 166
    electrical conductance of, 163–165
    formation of colour centres in,
        160–163
Solvation of electrons, 61, 99, 104
Somatic inheritance, 190
Sources
    alpha-, 20, 21
    beta-, 19, 20
    external, 13–26
    gamma-, 15–19
    labyrinth, 17–19
    machine, 21–26
    nuclear reactors, 21
    radioactive isotopes, 13–21
    spent reactor fuel, 21
    well-type, 17, 19, 21
Specific gamma-ray constants, 94
Specific energy loss, 36
Specific ionisation, 36
Spent reactor fuel, 21
Split energy levels, 73, 74
Spurs, 52, 103
Static electricity discharge, 20
Sterilisation, 183, 190
Stern–Volmer reactions, 72
Stopping power, 36
    mass, 86
Stroboscopic pulse radiolysis, 76
Strontium/yttrium-90, 189
Styrene, 175, 177
Subexcitation electrons, 61
Sulphonation of hydrocarbons, 184
Sulphonyl chloride, 183
Sulphur amino acids, 193
Synchrocyclotron, 26
Synchrotron, 26
System of units, 4–7
Szilard–Chalmers process, 27, 43

Target theory, 192

p-Terphenyl, 174
Tertiary structure, 191
Thermal electrons, 56
Thermal methods, 1
Thermal neutrons, 43
Thermal spikes, 157
Thermoluminescence, 168
Thermoluminescent dosimeters, 93
Thimble ionisation chambers, 85,
    87, 97
Thiocyanate dosimeter, 92
Third-body reactions, 60, 78
Three-body attachment, 60
Threshold energy, 42, 157
Topochemical reactions, 168
Tracks, 103
Trapped radicals, 181
Trapping centres, 165
Transient species, 28
Triplet state, 64, 65, 67, 79
Tungsten target, 23

Ultrasonics, 2
Units, 4–7

Valence bands, 163, 164
Van de Graaff machines, 23, 24
V-centres, 161, 162

Water radiolysis
    excitation process, 101
    ionisation process, 100, 101
    mechanism, 101–105
Water vapour, 150
Well-type source, 17, 19, 21
Wigner energy, 166, 167

X-ray machines, 11, 22, 23
X-rays, 2, 8, 9, 20, 22

Yield–dose graphs, 117